D0142951

Productivity in Natural Resource Industries

Improvement through Innovation

Edited by
R. David Simpson

Resources for the Future
Washington, DC

Printed in the United States of America

Published by Resources for the Future
1616 P Street, NW, Washington, DC 20036–1400

Library of Congress Cataloging-in-Publication Data

Productivity in natural resource industries : improvement through innovation / edited by R. David Simpson
p. cm.
Includes bibliographical references and index.
ISBN 0–915707–99–3

1. Coal trade—United States. 2. Petroleum industry and trade—United States. 3. Copper industry and trade—United States. 4. Forest products industry—United States. 5. Industrial productivity—United States. 6. Technological innovations—United States. I. Simpson, Ralph David.
HD9545.P76 1999
338.2′0973—dc21 99–21300
CIP

The paper in this book meets the guidelines for permanence and durability of the Committee on Production Guidelines for Book Longevity of the Council on Library Resources.

This book is a product of the Division of Energy and Natural Resources at Resources for the Future, Michael A. Toman, director. It was copyedited and typeset by Betsy Kulamer; its cover was designed by AURAS Design.

About Resources for the Future

Resources for the Future is an independent, nonprofit organization engaged in research and public education with issues concerning natural resources and the environment. Established in 1952, RFF provides knowledge that will help people to make better decisions about the conservation and use of such resources and the preservation of environmental quality.

RFF has pioneered the extension and sharpening of methods of economic analysis to meet the special needs of the fields of natural resources and the environment. Its scholars analyze issues involving forests, water, energy, minerals, transportation, sustainable development, and air pollution. They also examine, from the perspectives of economics and other disciplines, such topics as government regulation, risk, ecosystems and biodiversity, climate, Superfund, technology, and outer space.

Through the work of its scholars, RFF provides independent analysis to decisionmakers and the public. It publishes the findings of their research as books and in other formats, and communicates their work through conferences, seminars, workshops, and briefings. In serving as a source of new ideas and as an honest broker on matters of policy and governance, RFF is committed to elevating the public debate about natural resources and the environment.

Contents

Foreword

The crucial question, then, is not simply how much there is, but what it will cost to produce needed resource materials and services most efficiently... The threat of scarcity has been held in check, largely by technological forces.

Hans Landsberg, Leonard Fischman, and Joseph Fisher,
Resources in America's Future (Resources for the Future, 1963), p. 8.

Concern with natural resource scarcity, its economic consequences, and the prospects for ameliorating it have long been central issues for natural resource and environmental economists (and for Resources for the Future). Equally central has been the tenet that technological progress can indeed provide the means for alleviating the adverse effects of resource scarcity through increased productivity in the resource industries as well as through the development of new products.

Much of the literature on resource scarcity and technological advance, especially the more rarified theoretical inquiries starting in the 1970s, lacks specificity as to *what* technological innovation in the resource sectors amounted to, *how* it worked in practice, and *why* it occurred. For the argument that technological advance is an antidote to scarcity to be grounded in fact rather than supposition, it is important to go behind the theoretical artifices of shifting production frontiers or the calculation of statistical residuals in national income accounts.

The analyses in this book make an important contribution toward filling this gap in knowledge. From the case studies of individual sectors, we obtain rich descriptions of the nature of specific technological innovations and their sources. From the case studies and the statistical assessment of overall productivity growth patterns in these sectors, we obtain an up-to-date picture of broader trends that includes the impacts of many other factors as well as specific technological innovations. The introductory chapter by David Simpson, the editor of the volume, provides a clear summary of these findings and puts them in the context of wider concerns about technology, growth, and sustainability.

The book crystallizes useful insights about the process of technological innovation, both generally and in the sectors studied. Technological innovation arises from a number of causes: cost-increasing resource depletion, global competition (contrary to the Schumpeterian hypothesis of monopoly as the engine of innovation), and the agglomeration of many independent but complementary advances in knowledge and technique (thus adding weight to one of the principal premises of contemporary economic growth theory). The resource sectors examined here demonstrate a degree of technological innovation that will be surprising to readers who may have thought them antiquated. However, the causes and patterns of technological advance vary considerably from sector to sector, and overall productivity trends in the sectors reflect a mosaic of positive and negative influences of which specific technological advances are only one part. Facile generalizations about how technological innovation works, and how it contributes to economic advance in the face of natural resource scarcity, are not in order.

On behalf of Resources for the Future and all the authors in this book, I would like to express our deep appreciation to the Sloan Foundation for its support of this project. I would also like to express our deep appreciation to Jesse Ausubel, who worked hard and closely with both the Sloan Foundation and RFF in the structuring and execution of the project. A vote of thanks is due to the many participants in a March 1997 workshop convened by RFF to review the initial findings of the project. A great many specific individuals made significant contributions to the chapters in this book; these contributions are acknowledged separately in the chapters. I would further like to congratulate Chris Kelaher, Betsy Kulamer, and my other RFF colleagues responsible for the production of this volume.

Last, but certainly not least, I want to acknowledge one set of strong shoulders on which the other authors of this volume stand. For forty years, Hans Landsberg has been a towering figure in the fields of energy and resource economics and at RFF. Through his own contributions to the study of energy and mineral resources and technologies (including his contribution to this volume) and his direction of the work of others, he has given the rest of us a trove of both knowledge and inspiration. Hans, with the greatest respect and affection, we dedicate this book to you.

MICHAEL A. TOMAN
Director, Energy and Natural Resources Division
Resources for the Future

Productivity in Natural Resource Industries

1

Introduction

Technological Innovation in Natural Resource Industries

R. David Simpson

This book is about technological innovation in the extraction of natural resources and the effects of such innovation on U.S. natural resource industries. It is something of a paradox that the importance of technological innovation is underscored by the relative *unimportance* of these industries in national income statistics. Of a 1996 gross domestic product (GDP) of some $7.6 trillion, only about $240 billion, or 3.2%, originated in farming, fishing, forestry, and mining (CEA 1998).

This lack of statistical importance appears to belie two important facts. The first is that natural resource industries are essential. We would not have food to eat, clothing to wear, homes to live in, nor any of the other manufactured products that make up a far greater share of our economy, absent the raw materials these industries provide. The second fact is the sheer enormity of natural resource use. American forests produce some 400 million cubic meters of wood annually; if this total were assembled in a single block, it would measure nearly half a mile per side. The magnitudes are similarly astounding for other resources. Americans consume about twenty-five barrels—over 1,000 gallons—of oil per person per year. For all 270 million Americans, visualize this total by thinking of a lake of oil a mile across and half a mile deep. Americans consume about four tons of coal per person per year and nearly 700 pounds of metals per person per year. The physical quantities of natural resources produced and sold are enormous.

R. DAVID SIMPSON is a fellow at Resources for the Future.

Economic significance is measured by the product of physical quantities and prices.[1] Resource industries are not prominent in national income calculations because the prices at which these resources sell remain modest. Economic value is determined by scarcity, and the prices at which natural resources sell indicate that they are not, by and large, scarce. Predictions made a generation ago that by now we would be suffering through a period of resource scarcity and consequent impoverishment have failed to materialize. Food, materials, and energy remain relatively cheap. Consider Figure 1-1; the real price of natural resources[2] shows a declining trend. The price of these materials has declined by some 40% in the past forty years—although, as the figure demonstrates, this decline has not been uniform.

These price declines, in turn, are related to reductions in the cost of production. In Chapter 6 of this volume, Ian Parry presents statistics on the productivity performance of particular natural resource industries, but we anticipate his more detailed analysis with Figure 1-2. This graph shows how the quantities of at least one important and easily measured input—labor—devoted to the production of one natural resource has declined even as total physical production has increased in most instances. Output per worker has nearly tripled since World War II. As we will see, this finding is echoed in results for finer disaggregations of industries and broader classifications of productive inputs.[3]

In short, then, the output of the natural resource industries does not comprise a larger share of the national economy because the prices of these essential inputs remain relatively low. Prices remain low because these resources remain relatively abundant. These resources remain relatively abundant—even though their most easily accessed deposits have been depleted over time—because costs of production have not increased. Finally, costs of production have not increased because the inevitable effects of depletion have, to date, been more than offset by improvements in technology.

The task that we have set for ourselves in this volume is to better explain why it is that technology has been effective in controlling costs in the U.S. natural resource industries. There appear to be three factors explaining the technological progress we observe in these industries. The first is simply that necessity is the mother of invention. The most easily accessible reserves of natural resources have been depleted over time. As a consequence, costs of extraction would increase absent investments in cost-reducing technologies.

The second consideration in explaining innovation is that technological breakthroughs are ideas whose time has come. Innovation is an incremental and cumulative process. New machinery and processes are rarely truly novel, consisting rather of recombinations of existing technologies.

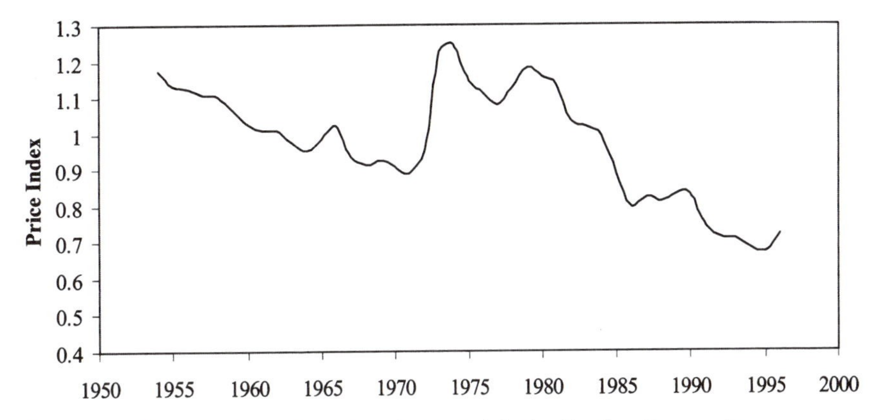

Figure 1-1. Price Index of Total Crude Materials for Further Processing.

Source: CEA 1998.

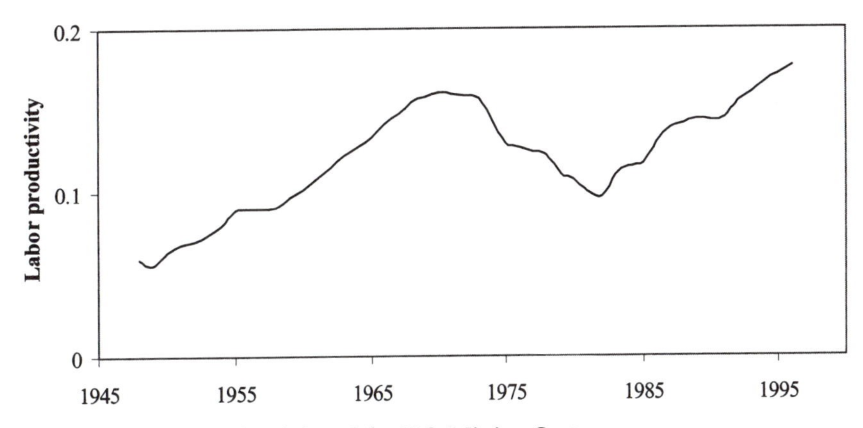

Figure 1-2. Labor Productivity of the U.S. Mining Sector.

Source: CEA 1998.

Moreover, a new technology seldom transforms an industry overnight. The process of technology adoption is typically gradual, with improvements in capability and expansion of applicability resulting from synergies with other technologies. While the depletion of resources may motivate experimentation, the general state of technology determines the set of possible outcomes.

The evidence that we have collected in this book provides many examples of depletion inducing innovation and of the combination and recombination of technologies producing still more innovations. There remains the third factor, which involves an ongoing ability to generate innovations. Some other countries supplying resources to world markets had greater initial resource stocks or have experienced less depletion than have U.S. producers. Yet the United States has managed to remain competitive in world resource markets even though geology and a longer history of depletion would seem to place it at a cost disadvantage relative to many other nations. It is true that, in some instances, the scale of U.S. production has declined in absolute terms or relative to world production. The fact remains, however, that U.S. firms are able to produce at costs that make them competitive with foreign rivals. This must be ascribed to an ability to develop and adopt new technologies more readily than can many international competitors.

This introductory chapter is followed by four industry case studies and a final chapter on the productivity performance of the four industries. In the following section of this chapter, we introduce the industries. Then we discuss and summarize the calculations of productivity reported in Chapter 6. We turn next to the three factors discussed above. The fourth section of the introduction describes the ways in which depletion has motivated innovation in the industry case studies. We then review the incremental and cumulative nature of innovation in our industries. In the final section of this introduction we review the evidence from the case studies concerning market, regulatory, and other conditions that may facilitate or impede innovation.

THE INDUSTRY CASE STUDIES

We have taken two approaches in our work. The first involves detailed case studies of important technological innovations in four U.S. natural resource industries. The four chapters that follow this introduction are, respectively, studies of innovation in coal mining, by Joel Darmstadter; in oil and gas exploration, by Douglas Bohi; in copper mining, by John Tilton and Hans Landsberg; and in forestry, by Roger Sedjo. The second approach consists of a statistical analysis of productivity trends in these four industries. The results of this analysis are reported in the final chapter of this volume, written by Ian Parry. In short, then, we have taken both bottom-up and top-down views of innovation in natural resource industries, with the former characterized by detailed examination of particular developments and the latter consisting of an aggregate summary of performance in the industries.

Coal

Several common themes emerge from the case studies. One is that resource industries, despite their long histories, are by no means technologically antiquated. The extraction or harvesting of resources from the land is among the most ancient of human activities. Yet the conditions under which these tasks are now accomplished in the advanced industrial countries bear little resemblance to the practices of the last century or, in some instances, even the last decade.

This point is perhaps made most clearly in Joel Darmstadter's analysis of coal mining in Chapter 2. Coal, a key resource in U.S. industrial development, continues to play a significant role in the economy due to its importance in electricity generation. An impression of the coal industry based on old photographs of grimy miners toiling under dangerous conditions is very misleading. The modern coal industry makes wide use of advanced technologies and has markedly reduced its labor intensity.

The coal industry also illustrates another theme common to our case studies. Technological innovation is not the only source of productivity change in the coal industry. In coal, as in some of the other industries we have studied, labor relations have significantly affected production. Some of the industry's poor performance in the 1970s may be ascribed to stormy relations between union labor and mine owners, and some of the subsequent improvement in productivity may be due to improvement in labor relations. Mine safety concerns were partly responsible for the labor unrest. The 1970s saw the passage of stronger mine safety and health regulations, which may have resulted in some reduction in measured productivity as well.[4]

In addition to safety and health regulation, the coal mining industry was also affected by environmental regulations passed in the same period. Restrictions on air emissions in the burning of coal had profound effects on industry structure. A large-scale shift occurred from Eastern, relatively high-sulfur, to Western, relatively low-sulfur coal. This geographical shift was also facilitated by new options for rail transport of coal from West to East. As Darmstadter shows, all of these factors combined with concurrent changes in technology to transform the domestic coal industry. While these effects of regulation are perhaps greatest in the coal industry, all of the industries we study have been affected to some degree by changes in regulation, and these changes have influenced the shape and pace of technological innovation.

Petroleum Exploration and Development

In choosing our industry case studies, we sought to compare and contrast different industries facing different circumstances. One important way in which the industries we studied differ is with respect to their resource reserves. Coal is abundant in many areas. This is in contrast to oil and gas, where decades of extraction have resulted in the depletion of the more easily accessible reserves in the United States. Douglas Bohi's study of exploration and development in the U.S. petroleum industry in Chapter 3 highlights the role of advanced technologies in the identification of new fields. Here again, the reader may be surprised to discover the sophistication of the technologies employed. Oil prospectors now are able to map geological formations hidden beneath a mile of rock with near-photographic precision. In some respects, this seems a more remarkable accomplishment than that of mapping distant galaxies.

Bohi shows, however, that the improvement in petroleum exploration and discovery cannot be tied to a single innovation. In fact, the main point of his chapter may be that innovations do not occur in isolation and that one technological advance is often augmented by others, with mutually reinforcing effects. This is another theme that comes up in all of our case studies, but it is perhaps most clearly illustrated in petroleum exploration and development. Breakthroughs in computer technology made new imaging technologies economically viable. Moreover, imaging technologies were more valuable as a result of concurrent developments in drilling technology and offshore production. Each technology enhances the effectiveness of the others. We will see this pattern of adoption of technologies from outside the particular resource industry, as well as mutual reinforcement of new technologies within industries, in all of the case studies.

Copper

In Chapter 4, John Tilton and Hans Landsberg's analysis of innovation in the copper industry makes an interesting contrast to the studies of both the coal and the petroleum industries. Unlike the case of coal, grave doubts were expressed even relatively recently concerning the survival of the U.S. copper industry. While one might expect the depletion of easily accessible deposits to result in an increase in production costs, the key to the viability of the U.S. copper industry appears not to have been the emulation of the exploration and development technologies that maintain the U.S. petroleum industry. In fact, Tilton and Landsberg show that U.S. copper producers have had at least one source of workable reserves lying almost literally under their noses, waiting for the advent of a technology to exploit it. The perfection of the solvent extraction–electrowinning (SX-EW) process enabled producers to recover commercial quantities of copper from waste dumps that had been abandoned after processing by earlier, less-efficient methods. The adoption of new technologies and the restructuring of copper firms have given a new lease on life to an industry once thought to be on its last legs.

Again, the findings in Chapter 4 are echoed in the experiences of other industries. In all of the industries we have studied, the depletion of the most easily accessible reserves has left producers with a choice between relocating extraction to other areas—or countries—or working their current sites more intensively. The latter option has been chosen to some degree in all four industries. Moreover, technological innovation has been an important factor in allowing producers to continue their existing operations.

Another important theme is raised in Chapter 4: the changes experienced in the copper industry were far from painless. Thousands of jobs were lost, mines closed, and companies were sold or went out of business. Many of these events were the consequence of increased competition from foreign suppliers. Despite these consequences of restructuring, Tilton and Landsberg argue that, in the final analysis, the U.S. industry has actually been made stronger as a result of being denied the protection from foreign competition that it requested in the 1970s. Those firms that survived were forced to innovate. As a result, the U.S. industry is arguably more competitive now than are many of its foreign rivals, who, despite the advantage of richer reserves, have not made the same investments in modernization.

While issues of international competitiveness are perhaps most conspicuous in the chapter on the copper industry, this is another theme on which all the case studies touch. The role of competition—or its absence—in motivating innovation has been an important theme of the industrial economics literature for decades. As will be seen throughout this book, competition has been an important element in spurring productivity growth.

Forestry

In contrast to the other three case studies, Roger Sedjo's discussion of forestry in Chapter 5 focuses on a renewable resource. It is ironic, then, that the consequence of depleting easily accessible resource stocks may be greatest in an industry in which stocks can be regenerated over time. Two broad factors constrain the growth of the U.S. forest products industry. First, many of the forests closest to centers of population and industry have been cut. The depletion of accessible stocks was accelerated by improvements in harvesting technology, which make it possible to eliminate large swathes of forest quickly. With the most easily accessible, easily harvested forests cut, it is more costly to harvest from the remote areas that remain. Second, among the industries we have studied, the forest products industry has perhaps been the most severely affected by environmental and other regulation. Forestlands tend also to be desired for recreational use or biodiversity preservation and are often withdrawn from production.

When it becomes more difficult to expand activities to new areas, production occurs more intensively in existing areas. In this regard, Sedjo argues that plantation forestry represents the wave of the future. Trees are increasingly coming to be treated as "crops" in the same sense as are food and fiber products. Planting increasingly occurs on the lands best suited for commercial forestry, as opposed simply to the areas in which trees had been growing. Cultivation of the crop now includes the application of pesticides, fertilization, irrigation, and preharvest thinning. Selective breeding and more advanced methods from biotechnology are becoming increasingly common in forestry. Improvements in trees themselves are becoming important, as is manipulating the environment in which the trees grow.

PRODUCTIVITY IN THE NATURAL RESOURCE INDUSTRIES

Each of the four case studies provides a narrative description of the genesis and diffusion of important technologies. In order to understand the role of these and other technological improvements in overall industry performance, however, we have also taken a more quantitative approach. In Chapter 6, Ian Parry reports on the overall productivity performance of the industries.

The contribution of Parry's chapter may be better understood if we first digress briefly on nomenclature. We are overdue in defining some terms. The words "technology," "innovation," and "productivity" can be difficult to define concisely. At a very broad level, we might say that each term is related to the extraction, manufacture, or, still more generally, transformation of goods. *Technology* might be defined as any aspect of the process by which goods are extracted, manufactured, or otherwise transformed. *Innovation* might be defined as any new way of extracting, manufacturing, or transforming.[5] *Productivity* means a measure of how much effort is required to extract, manufacture, or transform goods.

These definitions beg questions, however. Exactly how broadly should the terms be interpreted? While most people would agree that the invention and application of new types of machines are examples of technological innovation, what about new management practices? Government regulations? Labor relations? The focus of the following case studies is largely on technological innovation in the narrow sense of changes in the design, configuration, or capabilities of physical production processes. Some consideration of innovation in a broader sense is also important, however, for two reasons. The first is that management practices, regulations, and labor relations all interact with, respond to, and sometimes motivate changes in the design, configuration, or capabilities of physical processes. The case studies demonstrate that a discussion of technological innovation in the narrow sense could not be complete without some consideration of the environment into which innovations were introduced.

The Measurement of Productivity

The second reason for which we need to think about innovation more broadly concerns measurement, and here we return to the subject of Parry's chapter. He attempts to measure the impact of technological innovation on the performance of U.S. natural resource industries. It can, however, be extremely difficult to quantify the effects of better technology on the extraction of natural resources or, for that matter, the production of any good. A variety of methods have been proposed for measuring technological innovation. Patent applications filed and received, expenditures on research and development, and other measures of innovative effort have been proposed and implemented in empirical work. Inasmuch as our concern is more with the *effects* of innovation on the natural resource industries than with the *efforts* undertaken to achieve them, however, a measure of innovative *output* is more appropriate than one of innovative *input*.

For this reason, we have focused on productivity. Parry describes the calculation of productivity in greater detail, but it may be helpful to briefly review this important concept before presenting the case studies. *Productivity* is a measure of how much output one can generate using a given amount of input. Single-factor productivity indices are simple quotients: divide total output by total labor hours worked, for example, and the result is labor productivity.

Most of the case studies make some mention of labor productivity and, in some instances, emphasize it over other productivity measures. In a sense, labor productivity is what we *should* care most about. If in the final analysis we are concerned with human well-being, one of the things we might most want to know is how much output a worker makes per hour. Over the entire economy, one person consumes what another person makes, and labor productivity provides an aggregate measure of how consumption relates to time spent working.

There is, however, a problem with single-factor measures. This problem is exacerbated to the extent that we are measuring labor input and productive output in only a single segment of the economy. Labor productivity ignores the effects of other inputs. Labor productivity may go up solely because more machinery, or more material input, is employed. A better measure is *multifactor* or *total factor* productivity. Total factor productivity is computed by dividing output by the sum of all factor inputs.

Of course, one faces a problem of adding apples and oranges when summing over all factor inputs. Happily, elementary economic theory provides some help with this calculation. As Parry demonstrates in Chapter 6, consistent aggregates can be generated by weighting each factor by the share of payments made to it relative to the total value of output.

Productivity statistics are often reported as growth[6] in, rather than the level of, productivity. Productivity growth is calculated in a straightforward fashion. The year-to-year percentage growth in industry output can be decomposed into explained and unexplained parts. The explained part is the weighted sum of percentage changes in input quantities, where the weight given to each input is the share of expenditure on it divided by the value of output. The unexplained component of growth is just the difference between the percentage change in output and the share-weighted sum of percentage changes in measured inputs. This residual term is referred to as *productivity growth.*

Other Unmeasured Factors

In a sense, all of the constituents of the residual are "innovations." They are period-to-period changes that are not reflected among measured quantities of purchased inputs. They are innovations in the literal sense of "new things" that have arisen in the industry. Put in another way, this unexplained residual is "a measure of our ignorance" (Abramovitz 1956), a leftover term that cannot be further reduced by assigning it to measurable components. It would be inappropriate to assume that the entire unexplained component of growth in output is to be ascribed to the impact of technological innovations. In fact, there are several other possible components of productivity growth.

In calculating productivity growth, there are actually two considerations that separate those effects that can be measured from those that are incorporated in the residual. The first is that the changes in the physical quantity of a factor must be measurable. If we cannot quantify and observe a scale of technological attainment, for example, we cannot include it in the weighted sum of changes in measured inputs. Recall, however, that the weights used in calculating the sum of explained changes are the shares of input value relative to the value of output. This means that information on input prices is also necessary if we are to include changes in a particular factor in the "explained" portion of changes in output. We might measure things such as changes in air quality around a particular manufacturing establishment or changes in managerial structure within an organization. This information would not be helpful, though, since we would still be faced with the problem of adding apples and oranges in computing their contribution to overall output growth. Thus, if prices cannot be observed for particular factors relevant to production, we cannot separate out their effects.

One of the important things that may affect output growth but for which price data does not exist is environmental quality. It may appear at first blush that environmental regulations necessarily reduce productivity growth. Regulations increase costs of production by requiring producers to use more machinery, labor, materials, or other costly inputs to make their products. This increased use of inputs results in a decline in productivity, as currently measured.

Better measurement might reverse this conclusion, however. The fact that, by and large, producers are not currently required to pay the true price of the environmental degradation they cause means that they are using too many environmental "inputs"; they are, for example, overusing the waste disposal services provided by air and water. If producers were forced to pay the full price of the environmental damages they cause, it might well be the case that their production would decline by less than would their emission of pollutants. In fact, under this broader interpretation, environmental regulation *should* be productivity-enhancing: enlightened policymakers would enact stricter regulation only when it would increase the properly measured efficiency of production.

Another factor that we might expect to be important in the productivity of natural resource industries is the state of resource reserves themselves. If, as one would expect, the most easily accessible deposits are exploited first, followed by deeper, more distant, or otherwise less-accessible reserves, we would expect that productivity would decline over time. Other things being equal, the more production has taken place, the fewer easily accessible reserves are available, and the fewer accessible reserves available, the greater must be the contributions of labor, capital, and other variable inputs if output is to be maintained.

In principle, one could measure the impediment depletion imposes on productivity growth in the natural resource industries (and such attempts have been made; see, for example, Lasserre and Ouellette 1991 and Young 1991). There are some difficulties in implementing this idea, however. It can be difficult to estimate exactly the state of remaining reserves. Historical estimates of remaining resources have been notoriously inaccurate and, generally, pessimistic (Tietenberg 1996). In addition, the theory of natural resource pricing differs in one important respect from that of the pricing of other commodities. In resource industries, prices depend not only on the costs of production, but also on a generally unobservable "user cost" associated with the depletion of reserves. While statistical efforts have been made to estimate this user cost (see, for instance, Halvorsen and Smith 1991; Farrow 1985; Miller and Upton 1985), it cannot be measured precisely in most circumstances. For these reasons, Parry has considered any effects of depletion to be components of the residual, rather than attempting to calculate them separately.

Another unobservable factor affecting the measurement of productivity is industry scale. Generally speaking, some firms are better situated for producing certain products than are others. When the demand for natural resources increases, less-advantaged firms are drawn into the industry. Because these firms are less favored with respect to their location, reserves, and, in some cases, technology, they tend to be less-efficient producers. Thus, in periods during which demand is high, we might expect productivity to appear lower, and conversely during periods in which demand and industry scale contract, measured productivity may increase.

It is also difficult to measure the productivity effects of what one might call *managerial innovations*. The reorganization of a firm to shorten lines of communication or clarify responsibility might have profound effects on costs and efficiency, but its effects on productivity might not be identified from industry data. Similar considerations may be relevant with respect to improved labor relations.

A final consideration concerns the effects of productivity-enhancing innovations generated outside of an industry that eventually adopts them. Many technological improvements make their way into the public domain without acquiring patent or other intellectual property protection.[7] They are, then, underpriced, and their true contribution to output growth is not fully reflected in changes in measured input use. Thus, when these innovations are adopted, the measured productivity of natural resource industries increases. In this case, however, we are failing to measure all inputs in calculating productivity. Thus, the simple fact that industry productivity has increased does not necessarily tell us anything about the wisdom of investments undertaken elsewhere in the economy (perhaps in the public sector) that generate these spillovers.

Productivity Performance in the Natural Resource Industries

The productivity performance of the natural resource industries mirrors that of the economy as a whole. For reasons that are still poorly understood, productivity growth in the U.S. economy slowed substantially in the 1970s. Figure 1-3 depicts U.S. productivity from 1949 through 1993. While the overall rate of productivity growth was about 1.25% per year, most of that growth was concentrated before and after the decade of the 1970s.

At the time, one of the candidate explanations offered for the productivity slowdown was that the depletion of natural resources was a brake on economic growth. Indeed, an inspection of trends within the resource

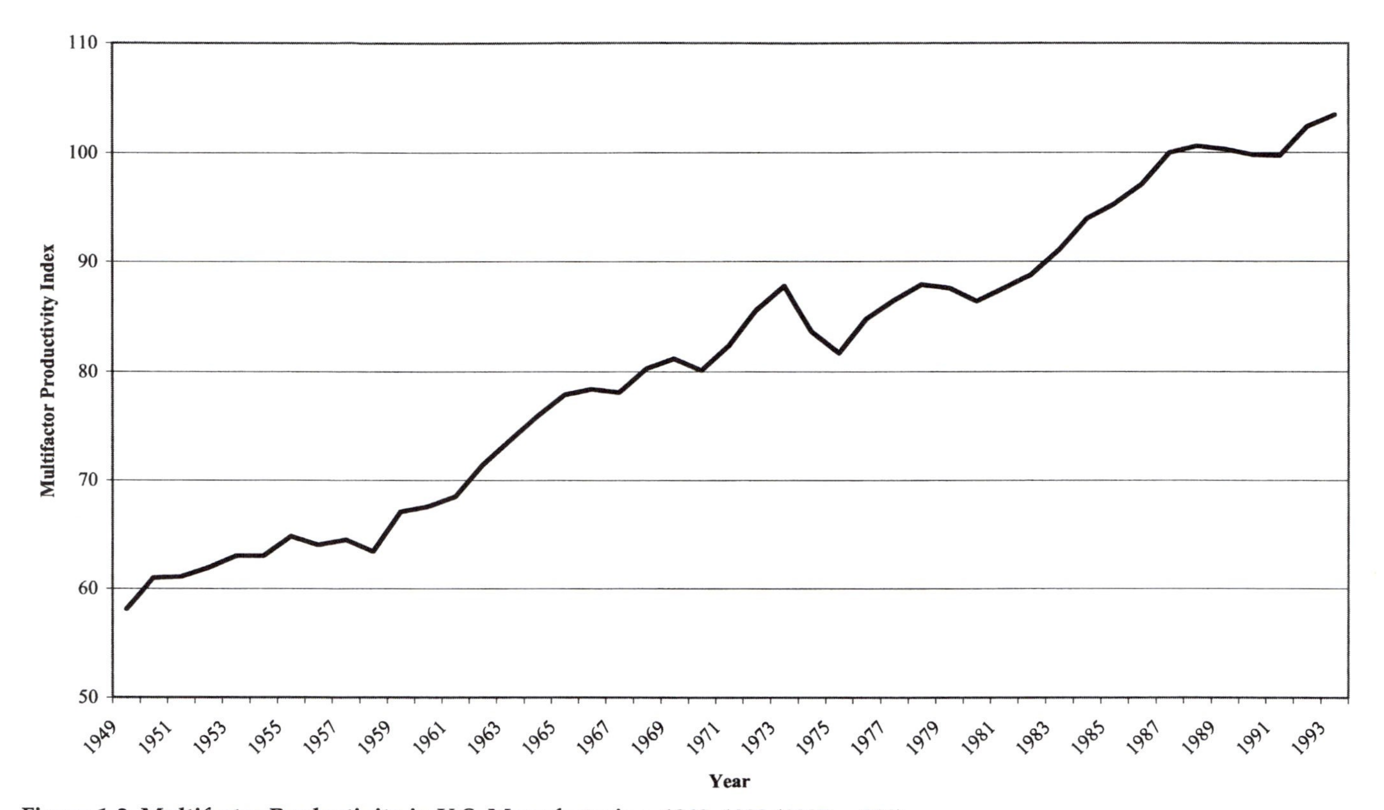

Figure 1-3. Multifactor Productivity in U.S. Manufacturing, 1949–1993 (1987 = 100).

Source: Bureau of Labor Statistics, http://stats.bls.gov/news.release/prod3.T06.htm (accessed March 1999).

industries themselves in the 1970s might have supported this view. As Parry shows, there were drastic declines in productivity during the 1970s in the four industries we have studied. If the depletion of resource stocks were constraining economic growth, however, one might expect the economywide productivity trend to have worsened over time, with the resource industries themselves showing the greatest productivity declines.

In fact, the resource industries seem to show almost the opposite trend. Following a decade of decline, Parry finds that productivity growth in the four industries improved considerably after 1980. In two of the four, the rate of productivity growth substantially exceeded that in the U.S. manufacturing sector as a whole.

While this recent experience provides some grounds for optimism, it is important to put the entire period in perspective. Based on Parry's statistics, overall productivity growth over the entire period of his sample (1970–1992) was close to zero in coal mining, substantially negative in petroleum and logging, and clearly positive only in copper mining. Thus, while the particular innovations discussed in the four case studies have surely had some impact in enhancing productivity growth, other factors have been at work as well.

A major component of the pattern is probably to be found in the history of U.S. resource industries and the international circumstances under which they operated. Due to the energy crisis and other factors, prices of resources generally rose in the 1970s and declined again afterward. The period of rising prices attracted the entry of less-efficient producers and expansion by existing producers into marginal areas. These developments resulted in declines in measured productivity growth: it takes more inputs to produce a barrel of oil or a ton of copper in an area in which reserves are less easily accessible than in an area in which they are more easily accessed.[8]

The 1980s saw a reversal of the price trends of the 1970s. Just as rising international prices had induced firms in the United States to enter and/or expand, they had the same effect on other producers around the world. This increasing international competitiveness drove prices down and eventually forced the less-efficient producers from the market. As less-efficient producers disappeared, measured productivity increased.

Parry notes that stricter environmental, health, and safety regulations may also explain part of the aggregate pattern. The enactment of tougher regulations forces producers to devote more inputs to the production of the same amount of output. This might lead to a once-and-for-all change in the *level* of measured productivity without affecting the long-run *rate* of year-to-year productivity growth. That is, tougher regulations might motivate changes in production technologies, but once these changes are made, the year-to-year pace of improvement might continue as before.

We have now reviewed the aggregate evidence on productivity growth and will shortly return to the lessons gleaned from the individual case studies. Before doing so, it may be helpful to summarize. While we are focusing on the role of technological innovation in the U.S. natural resource industries, it can be difficult to quantify the effects of such innovations. In particular, it may be difficult to separate the effects of those innovations we consider to be "technological" in nature from a host of other "innovations" in the form of changing regulatory, competitive, managerial, and other circumstances.

The record of productivity growth in the natural resource industries is mixed and consequently does not lend itself to easy interpretation. The most reasonable conclusion to be drawn seems to be that changes in market conditions and regulation had some temporary negative effects on productivity. Superimposed on these temporary phenomena may be the effects of the gradual depletion of more easily accessible reserves. Another long-run trend works in the opposite direction: the introduction and adoption of improved production technologies has offset the effects of depletion. It remains an open question as to which of these trends will eventually prevail.[9]

RESOURCE DEPLETION AND INNOVATIVE ACTIVITY

The observation with which we concluded the last section has been a subject of perennial interest in economics. As early as 1865, William Stanley Jevons, a famous economist of that era, voiced concern over whether Great Britain's coal reserves would be adequate to fuel that nation's continuing industrial expansion. Even at that time, the issue was not really new. An earlier generation of economists, represented most prominently by Thomas Malthus and David Ricardo, had devoted considerable thought to issues of resource scarcity. Malthus, in particular, is associated with the view that the expansion of human society is ultimately limited by the means of production.

In the twentieth century, similar concerns have continued to attract attention, although optimistic perspectives have been as much in evidence as pessimistic ones. The pessimists are perhaps most notably represented in *Limits to Growth* (Meadows and others 1972; see also Meadows and others 1992). The authors of that study concluded that humanity was using up natural resources faster than they could be replaced and, consequently, that disaster loomed.[10]

Other studies have reached more optimistic conclusions. Barnett and Morse (1963), in their study *Scarcity and Growth,* concluded that resource price and cost trends do not foreshadow any imminent problems (see also

Smith 1979). The analysis of cost trends in Barnett and Morse, in fact, can be interpreted as a sort of inverse measure of productivity.[11] Thus, their finding that costs did not increase as easily accessible resource stocks were depleted foreshadows our results. More recently, Nordhaus (1992) has drawn similar conclusions.

Implicit in any consideration of the ultimate effects of resource depletion are assumptions on technological change. In the past, potential crises were often forestalled by technological developments that enabled people to substitute one resource for another, find new reserves, and stretch limited reserves. If Jevons could see the world today, he might be shocked to find that the consequences of coal depletion he predicted have been obviated by a number of factors. We now rely much more heavily on other fossil fuels, new methods of production use resources much more efficiently, and state-of-the art methods enable modern miners to extract resources from depths that boggle the imagination even today and that would have been completely incredible to a writer in the mid-nineteenth century.

In fact, it is probably accurate to say that few economists regard the scarcity of extractable resources as a constraint that is likely to affect economic possibilities even well into the twenty-first century.[12] One of the reasons for this view, however, is that reserves have been found around the world. Thus, even though resource depletion does not appear to be a constraint on global economic possibilities, it has had an impact on particular industries in particular nations. If Jevons might be surprised to find Britain's energy needs met by other sources, he might still feel somewhat vindicated in noting that Britain's coal industry has experienced the decline he predicted.

It would be misleading to suggest that the substitution of distant for local sources of supply alone has obviated the link between technological progress and resource scarcity. Relocation of production is also, in part, a consequence of technological developments. Oil in the Arabian Peninsula, coal in the Powder River Basin of Wyoming, copper ore in Chile, and hardwood in Brazil would all be of little benefit to U.S. consumers if the technologies of modern transportation did not exist. Moreover, the extraction of resources from the less-hospitable environments of the world depends on the deployment of advanced technologies. In some instances, these are the ancillary technologies of transportation, and in some cases, they are the same technologies of extraction that are extending reserve life in the United States.

In short, then, a detailed examination of the effects of technology on resource extraction capabilities could reveal something about the ultimate limitations that resource scarcity places on economic growth. Addressing such broad questions is beyond the intended scope of our case studies.

For present purposes, however, the question might be turned around. Rather than asking how effective technology has been in keeping resource scarcity at bay, let us ask next how the depletion of easily accessible resources has motivated technological innovation.

Innovation as Intensification

As wells, mines, or forests are depleted, producers and consumers face choices. If production declines, prices will increase and consumers will shift to substitute products. At the same time, producers will look to other means to maintain supply. There are essentially two strategies for increasing supply. The depletion of easily accessible resources can motivate *extensive* or *intensive* strategies for maintaining production. Extensive expansion is the process of moving on to virgin prospects. It is essentially a notion of replication: whatever was being done in area A can simply be reproduced in area B—so long, that is, as there remain new areas with resources of comparable quality. Moving from one area to another may be expensive because the investments made in one area are not portable. In such situations, it may be wiser to make incremental investments in areas of existing operations than to replicate the entire package of assets elsewhere. This additional investment in existing operations represents an intensification of production.

One means of intensification is innovation. The application of better equipment can offset the effects of depletion. In fact, there is considerable evidence in the case studies that the depletion of easily accessible reserves has occasioned technological progress. Perhaps the best example of necessity mothering invention among our case studies is found in the solvent extraction–electrowinning (SX-EW) process in copper mining. U.S. companies faced difficult circumstances in the 1970s. They needed to reduce costs if they were to survive. This could be accomplished by relocating operations to other parts of the country or the world or by increasing the intensity with which mining companies worked resources at existing sites. The development of the SX-EW technology was an important component of the latter strategy. It enabled copper companies to "mine" the waste streams from their earlier operations; they could extract enough copper from mine tailings to make them viable ore sources.[13]

Bohi notes a similar phenomenon in the oil and gas industry. The U.S. petroleum industry faced a squeeze between competition from low-cost foreign producers and the upward pressure exerted on costs by the depletion of easily accessible domestic reserves. Under these conditions, it was imperative to develop techniques that would allow exploitation of known reserves at competitive costs. Initial extraction had removed as little as 30% of the oil in some abandoned reservoirs. This was largely because traditional vertical drilling methods limited the amount of oil that could be economically removed from reserves with complex structures. With the development of horizontal drilling, it became possible to approach a reservoir from any angle and thus to drain it more thoroughly.

The spread of plantation forests that Sedjo describes in his chapter is another example of depletion driving innovation. U.S. producers had often moved on to virgin forests after cutting the trees in one area. This practice has become less feasible in recent years, however. Remaining areas of virgin forest are often remote and inaccessible, and they can only be harvested at prohibitive expense. Moreover, the public is now more inclined to preserve remaining forests for recreation, wildlife habitat, and other nonconsumptive uses. Consequently, it has often proved more profitable in recent years to replant and manage harvested areas than to set out for the increasingly rare and inaccessible areas of virgin forest. While plantation forests are themselves "innovations," in the sense that they embody new management practices, they also incorporate biotechnological innovations. Investment in genetically improved trees yielded a limited return when extensive natural stocks remained to be harvested. As forestland is placed at more and more of a premium, however, investments undertaken to increase production per hectare and per year become increasingly attractive.

Of all the industries studied, coal may appear at first glance to be least affected by physical depletion. Darmstadter reports that identified reserves in the United States could meet domestic requirements far into the future. Having said this, however, it is worth adding that the quality of reserves has been a concern. The most easily mined deposits of coal have been exploited. The development of longwall mining resulted, at least in part, from the need to exploit more effectively thinner and deeper seams of coal. It is revealing to note that these techniques were pioneered in Western Europe, where declining coal reserves are a more important consideration than they have been historically in the United States.

THE CUMULATIVE NATURE OF INDUSTRIAL INNOVATION

While depletion drives innovation, innovation also perpetuates itself. Some of the major innovations that have transformed natural resource extraction borrow heavily from breakthroughs first made in other industries. Moreover, innovation typically does not proceed in a series of fits and starts, with infrequent spectacular breakthroughs punctuating long periods of business-as-usual. The process is, rather, one of gradual evolution. Even technologies subsequently recognized as revolutionary went through extended periods of adaptation and adoption. In many cases, additional technological developments were required to enhance the applicability of an initial innovation. It has also often been the case that one innovation does not achieve its full effectiveness until complementary, albeit ostensibly unrelated, technologies are developed.

These observations necessitate a clarification of our view of what motivates innovation. Depletion is certainly one consideration, but equally important factors are technologies available in the economy more generally and those being employed in the particular industries. The former provides a source of ideas for the borrowing, while the latter furnishes a set of assets to complement new developments.

Evolution of New Technologies in the Natural Resource Industries

These principles are best illustrated among our case studies in Chapter 3 on exploration and development in the petroleum industry. Bohi describes the physics of three-dimensional (3D) seismology, which is basically a process by which sound waves are used to map out the shape and location of underground geological formations. While the principles underlying the technique have been known for close to a century, its practical application had to await the development of high-speed parallel computing. Absent the ability to compile and interpret extremely large amounts of data, the technique would be far too slow and costly to apply in practice.

Today, 3D seismology is widely used in conjunction with directional drilling and in deepwater extraction operations. These three technologies are highly complementary. Having precise information about reservoir shape and location is less valuable if the technology is not available to enter a reserve from the optimal angle. This is precisely what directional drilling allows. Similarly, deepwater drilling is an extremely expensive process. It would not be economical were it not possible to obtain surveys sufficiently accurate to assure a high probability of success. This is what 3D seismology allows.

The chapter on oil and gas exploration is also instructive in that it shows how incremental progress has enhanced the applicability of new technologies. 3D seismology was initially used in the development of known reserves rather than in exploration for new ones. Since it was initially an extremely expensive process, it was thought to be worth the investment only if it were known in advance that there were in fact deposits to be mapped more accurately. As costs came down, however, it came to be applied to the search for new reserves as well as in the more accurate definition of known deposits.

Similar phenomena are observed in all four of the industries we studied. Consider, for example, what Sedjo found in plantation forestry. Again, complementary innovations had reinforcing effects. Expensive irrigation and fertilization systems, for example, make little sense unless one can be reasonably sure that the seedlings they water and feed will grow into commercially valuable trees. Conversely, selective breeding is conducted to develop trees with superior commercial traits, such as superior growth and wood quality. An unintended consequence of genetic selection for superior commercial traits may be reduced resistance to drought, disease, and pests. Thus, breeders can expect greater success when selecting trees for their commercial attributes to the extent that tree growth and survival can be enhanced by irrigation, fertilization, and pesticides.

The biotechnological advances now being adopted in forestry owe a great deal to discoveries made in medicine and agriculture. The principles of selective breeding were first developed in the improvement of annual agricultural crops and in animal husbandry. Medical and chemical researchers discovered the basics of molecular genetics. These techniques have been profitably applied in forestry, but this again is an instance in which innovations were adopted from other contexts, rather than developed within the industry.

Darmstadter's study of the coal industry affords additional examples of complementary technologies. The productivity advantage of longwall mining relative to the room-and-pillar method that was employed previously has increased in lockstep with the improvement in power and positioning technologies for deploying it. In fact, the advent of ever-larger and more powerful machinery has enhanced overall coal productivity. These developments depended on the generation of a host of mechanical improvements.

We have been reviewing instances in which industry-specific innovations spring from technological innovation more generally. In many cases, these more general developments are so pervasive as to be taken for granted. For example, mechanical power revolutionized both coal mining and forestry. Each industry benefited from the availability of increasingly compact and powerful engines. The pickax and the handsaw were replaced by, respectively, mining machines and chainsaws at roughly the same time as internal combustion engines were revolutionizing transportation and industry more generally. The increases in industrial productivity that high-speed computing is now generating may parallel the earlier impact of steam, electric, and internal combustion power.

In Chapter 4 on copper mining, Tilton and Landsberg also point to the ways in which the most basic of innovations come to revolutionize ostensibly unrelated industries. Copper ores have been leached since the fifteenth century. A variety of ancillary developments have made the process more efficient and economical. One of the most important of these is something that we now take for granted. Electrowinning requires electricity. Electrical generation and distribution systems make the SX-EW process possible.

Copper mining also provides evidence of the ways in which innovations are refined and improved to increase their efficiency and widen the scope of their application. The discussion in Chapter 4 is detailed, but we can give some idea of the range of developments by just noting their variety. Innovations as varied as the addition of cobalt salts to avoid lead contamination, use of bacteria in leaching, and development of water conservation methods for use in arid regions have been important in broadening the application of the SX-EW process. These are all examples of the way in which a basic idea is developed, extended, and combined with a wide variety of subsequent innovations to yield continuing improvements.

It is also interesting to note that the SX-EW process now used in copper mining was itself adopted from another industry. The process was first used in the wartime refining of uranium for weapons production. It has since been adopted for use in several metals industries. Spillovers of knowledge and experience have surely occurred between the different applications.

Tilton and Landsberg also identify a wide range of players involved in the evolution and diffusion of the SX-EW process in copper mining. When Ranchers Exploration and Development Company began operating the first commercial SX-EW facility for copper at its Bluebird Mine in Arizona in the late 1960s, it worked closely with the General Mills Corporation, Hazen Research Incorporated (an engineering consultant), and the Bechtel Corporation (a construction and engineering company that built the first processing plant). In some instances, then, the interindustry flow of information may be facilitated by the formal involvement of firms from different industries.[14]

There are, of course, instances in which the development of certain technologies defines watershed events in the evolution of an industry. To give one example, Bohi describes the history of oil and gas exploration as an episodic one. A technology is developed with which one type of deposit can be identified; deposits of that type are discovered and exploited; a technology is developed with which a second type of deposit can be identified; deposits of the second type are discovered and exploited; and so on.

Closer inspection of industry histories suggests that such episodic evolution may represent the exception that proves the rule. There is no denying that certain technologies have revolutionized the natural resource industries, but in each case these technologies were adopted more widely as they evolved and improved. Moreover, the clear evidence of all the case studies is that even major innovations are accompanied by a number of ancillary developments that enhance their efficiency and broaden their applicability.

CONDITIONS FOR CONTINUING INNOVATION

We have reviewed evidence concerning two factors that are important in inducing innovation. The first is simply the pressure of circumstances. The second concerns spillover benefits from complementary technologies. These considerations do not provide a complete explanation of how and why innovation occurs, however. The evidence from the case studies, as well as from more general data on world economic development, is that innovation among firms and countries tends to persist over time. Those who pioneered new technologies in one period are likely to do so again in the next. To some extent, this might be seen as a consequence of the factors we have already cited. If depletion induces innovation, further depletion may induce further innovation. And if the existence of one generation of technology creates conditions for the birth of another, the firms and countries that produce the first generation may be better positioned to produce the second.

It is not clear that one can infer from these arguments that technological advantage will persist, however. Perhaps one nation does deplete its reserves faster than does another, but eventually both nations will experience scarcity of readily accessible reserves. The question then becomes, will the initially more richly endowed nation also develop or adopt technologies to offset depletion? Similarly, perhaps it is the nation that first begins to run low on readily accessible reserves that first begins to innovate. But if innovations build on one another, what prevents one country from benefiting from innovations that another pioneered?

While there is considerable evidence of the international diffusion of innovations, there is also evidence that technological leadership tends to persist. In this final section, we discuss some factors that might explain this phenomenon in the natural resource industries. We begin by reviewing evidence on the role of industrial competition in innovation. Then we discuss the effects of regulation. Finally, we address the role of cultural factors.

Competition and Innovation

There is a paradox of innovation. On one hand, innovations are developed in order to cut costs and improve profitability. On the other hand, because one person can sometimes copy another's innovation, each might prefer to be a *free rider*. This means that you might prefer to wait for me to produce an innovation and then appropriate it for your own use, rather than to go to the expense and effort of developing it yourself and, in the process, providing it for me as well. Thus, a situation could develop in which everyone would profit from innovation, but no one would attempt it.

This paradox has fueled another long-running debate in economics. This debate has sometimes been characterized as one concerning the Schumpeterian hypothesis. The noted Austrian-American economist Joseph Schumpeter argued that monopoly is a precondition for innovation. His view has been interpreted in two ways. The first is that a preexisting monopolist is more likely to innovate than would be a competitive firm. The second interpretation is that the grant of a monopoly through the award of a patent is a necessary evil if innovation is to be encouraged (Tirole 1988).

With regard to the first of these interpretations, at least it remains an open question as to what market conditions most encourage innovation. Is progress more likely when large numbers of small competitors race to be the first to develop new technologies or when a single large entity devotes its considerable resources to research and development? If capital investment requirements are modest and inventions cannot easily be duplicated by rivals who were unsuccessful in developing their own innovations, competition would be most conducive to innovation. Conversely, if capital investment requirements are large and successful innovations can be easily copied by free riders, only a single large entity will have both the funds and the incentive to engage in innovation. In the extreme, some have argued that public bodies should be charged with conducting research.

There is, as yet, no decisive empirical answer as to how competition affects innovation in U.S. industry.[15] Such evidence as we can adduce from our case studies suggests that vigorous competition is, at the very least, not inconsistent with innovation. This view is borne out in a consideration of coal mining under different political regimes. Mines in Eastern Europe, the former Soviet Union, and China tend to lag behind those of the West in productivity. Conversely, those mines that remain operating in Britain and Germany are world leaders in mining technology. This is despite the fact that, in Darmstadter's words, Western European mines suffer from "the inexorable result of unfavorable geological conditions." While part of this leadership may be due simply to the necessity of employing cutting-edge technology to offset geological disadvantage, one might also conjecture that differences in performance can be ascribed in part to differences in the incentive systems in place until recently.

While complete public ownership might be expected to reduce incentives for innovation and efficiency, one can make a reasonable argument that basic research should be publicly funded: governments, to the extent that they can represent broad cross-sections of interests, will not be tempted to free ride. We can point to several instances in which public programs were important in the development of important new technologies. As noted above, the solvent extraction process now used in copper mining was first employed in mining uranium for military purposes. Darmstadter notes that global positioning systems are used in the coal industry to make transportation more efficient, and Bohi reports that satellite communications are used in transferring seismological data from petroleum exploration. These technologies are by-products of publicly funded space research. Similarly, some advances in the computer technologies that have revolutionized virtually all industries were publicly funded. Finally, public funding has played an important role in both practical programs of crop improvement and in basic research in molecular genetics. These programs have provided the foundation for genetic improvement programs in forestry.

It is also interesting to note, although admittedly somewhat difficult to classify, that public funding has also played a role in the *diffusion*, as opposed to solely the generation, of new technologies. Darmstadter has noted that British and German firms first developed longwall mining technologies, but that one of the factors associated with its introduction and improvement in the United States was support from the U.S. Bureau of Mines.

While connections can be drawn from public research to the private adoption of innovations, it is not clear how close the links have been, nor how essential public funding has been in underwriting basic research. In a world with so many intricate interconnections, it would be more remarkable if one could *not* demonstrate some link between publicly funded research and particular resource industry innovations. Moreover, industrial innovation tends to be cumulative and incremental. To say that some links in a chain of innovations were provided by public funding does not establish how crucial each was. Finally, one can never be sure that even the most basic of research, if it had not been publicly funded, would not have occurred otherwise.

With respect to applied research, the case studies by and large support the view that competition begets innovation. Darmstadter notes that firms must be large enough to mobilize the capital investment required to operate a mine of efficient scale. Yet he follows this observation immediately with evidence that the concentration among large firms in the coal industry is relatively low (by the standards of, say, the automobile, aerospace, or computer industries). Taken as a whole, therefore, the coal chapter provides no support for the thesis that market power is necessary for innovation. To the contrary, it provides ample evidence that a period in which regulatory and other considerations generated greater competition between Eastern and Western producers was also one in which tremendous technological strides were made.

In forestry, it has typically been the largest firms that have devoted the most effort to in-house genetic improvement programs. Weyerhauser and Westvaco have been among the leaders, an observation which would seem to be consistent with the view that they have substantial plantation forest areas over which to amortize their investment. Again, however, the innovative efforts of these large domestic firms may well have been motivated by the need to keep pace with abundant foreign competition.

Similarly, Bohi's chapter on oil and gas exploration notes that some of the world's largest firms have been pioneers in the deployment of new technologies, with Amoco and Exxon among the leaders in adoption of 3D seismology, Arco and British Petroleum in horizontal drilling, and Shell and Petrobras in deepwater systems. The first five are all large multinational organizations with global activities. Petrobras, as the oil company of one of the world's largest nations, may be the exception proving the rule that expensive innovations are conducted by firms large enough to use their results broadly.

Bohi also attributes the generation of superior technologies to difficult, as opposed to comfortable, market conditions. Domestic producers found themselves in the early 1980s in something of a double bind. Declining real prices for petroleum products constrained their profitability, while at the same time they faced strong competition from more geologically advantaged nations. Even under these difficult circumstances, domestic producers developed and expanded the use of new technologies that enabled their continued survival.

The connection between competition and innovative performance is drawn most forcefully in the study of the copper industry. Tilton and Landsberg concede that some short-run variations in other factors affected both the U.S. copper industry's decline and its subsequent recovery, but they argue that the long-run fact is that U.S. copper firms operate in a highly competitive international market. In their estimation, competitive forces were not extraneous or inimical to the modernization of the industry but, rather, drove it. Tilton and Landsberg go so far as to surmise that protection of U.S. copper firms from their international competitors would, in the long run, have hobbled their innovative activity and eventually threatened their survival.

The discussion in Chapter 5 of competition as a motivation for innovation in forestry has necessarily been compressed, as other issues demanded more space. Nevertheless, Sedjo identifies competition from foreign forest products as an important factor in the development of plantation forestry.

It is tempting to conclude that the case studies establish a causal connection between competition and innovation. This conclusion would be premature. While it is clear that major innovations have emerged in conjunction with vigorous competition, we do not have enough data to conclude that competition is essential to innovation. The fact that productivity gains occurred as global integration increased in the 1980s and 1990s comprises only circumstantial evidence. The case studies cannot tell us what would have happened if there had been less competition. What is apparent at the very least, however, is that considerable innovation has occurred in industries characterized by intense competition, both among firms within the country and between domestic and foreign firms.

Regulation and Innovation

We have mentioned the argument that basic research should be conducted in the public sector or, at least, subsidized with public funds. Evidence from the case studies, however, suggests that public policy may have had a greater impact via the inadvertent effects of ostensibly unrelated regulations than through attempts to induce innovation directly. Having said this, we should quickly point out that there are few broad generalizations one can make about the effects of government regulation on innovation. In fact, perhaps the only such generalization is that effects are far from uniform and may not even be easily predictable.

This should not be too surprising. Different regulations have been established to achieve very different ends. Moreover, the final form of a regulation can often reflect very different objectives than appear to have been embodied in the statute underlying it. Environmental regulations, for example, have sometimes contained provisions grandparenting existing polluters, mandating particular technologies, or otherwise favoring one segment of an industry over another.

The most obvious impacts are found when regulations specify the resources that can be exploited. In forestry, stricter regulation of land use and requirements for the maintenance of wildlife and public use areas have had a tremendous impact on harvest opportunities in the United States. Such policy-driven reductions in the resource base have had as much effect on encouraging innovation as has physical depletion.

Similar phenomena are noted in Darmstadter's study of the coal industry. In some instances, government regulation has taken the form of land use restrictions. Some areas are off-limits for coal mining, while in others stricter regulations have increased miners' expenses for land restoration. A more important set of restrictions concerns the regulation of coal itself. The Clean Air Act and its subsequent amendments have placed tight restrictions on sulfur dioxide emissions from burning coal and effectively, therefore, on the sulfur content of U.S. coal.[16] Environmental regulations played an important role in the pronounced shift from East to West in the location of coal production.

This westward movement in coal production raises questions. Darmstadter finds a geographical effect in coal industry productivity improvement: part of the improvement in measured productivity is not due to mines in general becoming more efficient but, rather, to more efficient mines accounting for a greater share of output. But if Western mines are generally more efficient, why have they not always dominated U.S. coal production? Darmstadter considers this question and offers a number of answers. Among them is the evolution of another type of regulation. The addition of more rail transport capacity from the Powder River Basin to Eastern coal markets is in part a consequence of federal regulators' diminishing willingness or ability to restrict competition in long-distance freight markets. Concurrently, the scale of Western surface-mining operations facilitated long-term contracts for transmitting "coal by wire," that is, electricity from minemouth generating plants. This process may be further encouraged by the ongoing deregulation of nationwide electricity markets. Finally, the easing of policy disputes over development of federal lands, on which most Western coal resides, stimulated the westward shift as well.

Tilton and Landsberg provide something of a different perspective on the effects of regulation. Environmental regulations, on occasion, can have unintentional benefits beyond those specifically targeted. For example, regulatory requirements to recover sulfur from smelter emissions increases the supply of sulfuric acid available for copper mines to use as a leaching agent in the SX-EW process. Moreover, Tilton and Landsberg raise the possibility that keeping pace with tougher regulation may increase the incentives to develop more efficient technologies. Thus, the unfavorable impact of regulation on productivity in the short run may be mitigated to some extent by the long-term benefits of the innovations regulation may induce.

As a final example of the effects of regulation, consider Sedjo's findings on plantation practices in the United States. Some jurisdictions treat crops harvested on relatively short rotations as "agricultural" rather than "forestry" products and subject them to less-stringent regulations. While incentives exist to shorten harvest cycles for other reasons, the desire for greater regulatory leniency also adds to the motivation for innovation.

Perhaps the only conclusions we can draw about the effects of regulation on innovation are, first, that some regulations have had profound effects but, second, that the direction or magnitude of such effects cannot be predicted in general.[17] Even so, there does seem to be another public-policy-related phenomenon that augments the long-term pace of innovation. The United States continues to produce or adopt innovations that, in turn, allow it to remain cost-competitive with nations that are more favorably endowed with physical resources. Perhaps something in American cultural circumstances favors continuing innovation. We turn to this conjecture in concluding this introduction.

The Cultural Prerequisites of Continuing Innovation

One of the factors we have found to motivate innovation is the necessity of lowering costs. This observation is hardly original, nor is it profound.[18] Those firms whose very survival is threatened by higher costs of production are under greater pressure to reduce costs than those whose costs are already low. The simple fact that different countries have different cost conditions begs other questions, however. If there is global competition in the provision of natural resource commodities, how can a nation's industry survive if it is at a cost disadvantage vis-à-vis its international rivals? Why would not the nation(s) that can supply resources at lowest cost do so, to the exclusion of others?

The answer is that cost advantages must be defined relative to circumstances. Constraints on supply may prevent even advantaged producers from supplying the entire market. If, beyond a certain level of production, a supplier's costs begin to increase rapidly, it no longer enjoys a cost advantage relative to its rivals. Moreover, cost advantage may be compounded of a variety of factors. One nation's industry may be better situated with regard to the resource deposits at its disposal, while another's may be favored with a more skilled labor force, better equipment, or superior transportation facilities.

But this raises another question. More abundant and accessible resources would imply a cost advantage *if other things were equal*—but why should other things *not* be equal? More skilled labor, better capital equipment, and other factors of production can be acquired by investment. Why do some countries that enjoy an advantage with respect to their endowments of natural resources fail to press that advantage by investment?[19]

We have, in the course of this introduction, touched on several unanswered questions in economics. We are concluding with what is perhaps the most important of them. Why is it that some nations invest and grow rich, while others do not?[20] Simple theory predicts that the return to capital investment[21] ought to be higher in areas in which the capital stock is low relative to quantities of labor, resource reserves, and other factors of production (see, for instance, Lucas 1990; Mankiw, Romer, and Weil 1992). The prediction of this "simple theory" is that investment would flow from the wealthier to the poorer nations of the world. This is manifestly not the case. In fact, it appears that the opposite is often true. Japan, for example, has relatively few natural resources but receives a great deal of capital investment. Many nations of Africa, on the other hand, have been richly endowed with mineral and other resources but receive very little capital investment.

A great many explanations for this apparent paradox have been proposed. Models have been constructed in which qualitative differences in labor or other inputs, differences in social and legal institutions, or the self-reinforcing advantages of proximity, by which growth begets growth, have been proposed as resolutions of the paradox (see, for example, Romer 1994; Grossman and Helpman 1994). While there is little evidence with which to distinguish among these theories at present, a common theme in recent literature is that capital investment and, particularly, investment in high-technology equipment either create or give evidence of the existence of conditions under which further investment in high-technology equipment is profitable.

We have already reviewed in some detail evidence concerning the incremental and cumulative nature of technological innovation and adoption in U.S. resource industries. We won't repeat that review now, but only point out that the processes of incremental improvement, borrowing related technologies, and recombining existing technologies to generate innovations are facilitated by corporate ties and physical proximity. Even in the absence of these factors, some degree of openness with respect to information sharing, as may occur among firms within and between advanced industrial countries, would appear to be conducive to innovation.

A more basic consideration yet is that innovators, if they are to have an incentive to innovate, must have some confidence that they are going to enjoy the rewards accruing to success. Many regions blessed with abundant resources lack what Sedjo characterizes as the "political, social, and cultural prerequisites" for world-class production. It would be a good thing for both humanitarian and pragmatic reasons[22] if developing nations could quickly acquire these "political, social, and cultural prerequisites." To the extent that they do not, however, we can anticipate that the U.S. forest product industries—and other natural resource industries—will continue to compete despite the disadvantages other circumstances may impose.

The existence of appropriate incentives is a prerequisite for innovation. Over and above this, however, there is reason to believe that those nations and the firms within them that develop or adopt state-of-the-art technologies are more likely to stay at the leading edge of innovation. Examples we have cited earlier suggest that the process of innovation can gather momentum and perpetuate itself, with one improvement building on another. It seems reasonable to suppose that those who have made a practice of adopting, experimenting with, and improving one new technology will be in a better position to do the same with another.

In his study of oil and gas exploration, Bohi points to the role of firms and countries with a willingness to experiment. While one should not underestimate the importance of physical conditions in inducing innovation—countries with no access to deepwater areas would, for example, have little incentive to invest in deepwater equipment—other factors are also at work. Success spawns imitation, but innovations would not appear absent the pioneers who first test them. Cutting-edge technologies are now readily available to oil and gas producers around the world. This does not negate the point concerning the importance of innovative leadership, however. The transfer of technology and expertise is conducted under contracts by which innovators are rewarded for their knowledge.

Tilton and Landsberg speculate that the United States may maintain a technological lead in copper production for the foreseeable future. There is an element of necessity-mothering-invention to this prediction—the United States did, after all, pioneer the SX-EW process in response to a decline in readily accessible reserves—but there may also be other forces at work. Tilton and Landsberg also note that those firms and nations that demonstrate "a persistent commitment to innovation and to the development and use of new technologies" are likely to remain profitable even if their new technologies can be copied in a matter of years or even months.

Inasmuch as the line between successive innovations becomes blurred by recombination and synergy, one might say that successful research and development have as much to do with continuing experimentation as they do with the search for specific breakthroughs. Those blessed by experience, temperament, and, perhaps, cultural support, who also have the willingness to experiment, may remain industry leaders for longer periods than their resource endowments suggest.

ACKNOWLEDGMENTS

I would like to thank the chapter authors, and especially John Tilton, for helpful suggestions for the improvement of this introduction. Michael Toman provided both careful proofreading and overarching vision. Comments from two anonymous reviewers and Jesse Ausubel are also gratefully acknowledged. Brian Kropp, Nancy Bergeron, Joyce Luh, and Pamela Jagger provided excellent research assistance. Remaining errors are my responsibility.

ENDNOTES

1. This is something of an oversimplification. It is actually the value added by industry, which is measured by the difference between revenues and expenses, that comprises the measure of sectoral contribution to GDP. Scarce resources command prices greater than their unit costs of production, however, so high prices do contribute to a large share in GDP.

2. We include in this category foodstuffs and feedstuffs, energy, and other "crude materials for further processing" (see CEA 1998).

3. We should also note that, in the industries we have studied, as with this broader aggregrate of resource industries, performance was particularly poor during the decade of the 1970s. Rebounds also occurred, but not always with a full recovery.

4. The issues here are unclear, however. On one hand, labor and total factor productivity might have fallen as a result of more time and effort being devoted to protecting workers as opposed to removing coal. Another response to personnel safety concerns may have been increased mechanization, however, and this might be expected to have increased labor, if not total factor, productivity.

5. In the industrial economics literature, distinctions are drawn based on Schumpeter's (1943) taxonomy. *Invention* is the creation of a new technological idea, *innovation* is its first commercial use, and *diffusion* is the spread of innovations across users. (We are grateful to an anonymous reviewer for suggesting this taxonomy.) In our context, however, it may be better to think of innovations in the broader sense we propose here.

6. It should be emphasized that productivity growth need not necessarily be positive. In fact, Parry reports that productivity growth was negative for all four industries in the decade of the 1970s.

7. Even patented products may not be fully protected against unauthorized appropriation. Simply demonstrating the feasibility of an idea may allow others to imitate it without infringing the patent.

8. We run into some semantic problems in discussing these trends. Resource stocks themselves can be considered inputs to the production of resource commodities; in other words, coal in the ground is an input that affects the amount of coal mined. If we had reliable measures of the stocks of resources available for exploitation, the "true measure" of productivity might not decline, as a reduction in output could be related to a corresponding reduction in the resource stock input. As it is very difficult to measure resource stocks, however, existing productivity measures may confound the effects of resource depletion with those of other factors that constitute the residual.

9. An anonymous reviewer has noted that neither trend may "prevail" in the long term; during some periods resource depletion will be dominant and, during others, technological innovation. Even to suppose that such periods will continue

to alternate ad infinitum, however, seems to presume that resource depletion will never have catastrophic implications.

10. We may now be entering another period of depletion concern; see, for example, Campbell and Laherrére 1998 and Kerr 1998.

11. The economic theory of duality can be invoked to derive productivity as a residual measure of cost reductions that cannot be explained by changes in input prices.

12. There is, however, more debate concerning the limits imposed by the environment. Some would argue that climate change, ozone depletion, biodiversity loss, and other factors may affect our well-being in more profound ways than would petroleum or mineral depletion.

13. It is interesting to note just how much U.S. copper companies have relied upon working existing mines more intensively, as opposed to opening new ones. Tilton and Landsberg report that annual U.S. copper production increased by more than 40% from the 1970s to the 1990s, yet few of the mines operating in 1995 were opened after the 1970s, and together these new mines accounted for only some 3% of U.S. copper production in 1995.

14. This is clearly not a panacea. Darmstadter notes that the ownership of coal properties by oil companies was not, in general, financially successful.

15. This is not, we might quickly add, for want of interest in the topic. One important strand in the recent literature has been a "new Schumpeterian" approach to innovation and growth; see, for example, Aghion and Howitt 1998.

16. Additional restrictions may be imposed as a result of the United Nations Conference on Environment and Development, which met in 1997 in Kyoto, Japan. As Darmstadter notes, policies enacted to achieve these reductions could have a tremendous impact on the coal industry.

17. It is also worth noting (as an anonymous reviewer has suggested to us) that the effects of regulation on innovative behavior can be subject to a type of "selection bias." Case studies concentrating on the appearance of important innovations may be able to identify those that occurred in response to regulatory pressures, but they cannot, in general, identify those innovations that did *not* occur because of regulatory pressures. Yet it is sometimes argued that the latter are important casualties of technology-, as opposed to performance-, based regulation.

18. It is, however, somewhat problematic. It is not obvious, as a matter of economics at least, that a firm would be more anxious to increase its profits from, say, nothing to $1,000 than it would be to increase them from $100,000 to $101,000. Here, perhaps, we might simply admit that the psychology of desperation can provide extraordinary motivation.

19. One possibility here is that the possibility of imminent exhaustion makes expensive investments in cost reduction unremunerative. Consider a hypothetical example. Suppose that a country has very small reserves of ore that can be mined at negligible cost. If no amount of additional investment could expand the volume of national reserves, investment would not occur. In such circumstances, however,

one would expect that this low-cost producer would be the first to exhaust its resources.

20. There is a literature on whether different countries will eventually converge to the same level of income, but it is both too vast and too inconclusive to cite in detail. The interested reader is referred to recent summaries by Baumol and others (1994), Jones (1997), and Pritchett (1997). It may be worth noting, however, that the issue of whether different countries will *eventually* converge to the same income level is of limited practical relevance if the rate at which the gap narrows—if indeed it does—is glacial.

21. Implicit in what we are saying here is that innovation is embodied in or, still more generally, is a form of capital investment.

22. One pragmatic reason for which more reliable social infrastructure in developing countries is desirable—even to the developed countries that would compete with them for the provision of natural resources—is that it would facilitate the licensing of technologies from developed to less-developed countries. An impediment to such licensing now is a perceived inability to enforce contracts calling for royalties.

REFERENCES

Abramovitz, Moses. 1956. Resource and Output Trends in the United States since 1870. *American Economic Review* 46(2, May): 5–23.

Aghion, Philippe, and Peter Howitt. 1998. *Endogenous Growth Theory.* Cambridge, Massachusetts: MIT Press.

Barnett, Harold J., and Chandler Morse. 1963. *Scarcity and Growth: The Economics of Natural Resource Availability.* Baltimore: Johns Hopkins University Press for Resources for the Future.

Baumol, William J., Richard R. Nelson, and Edward N. Wolff. 1994. *Convergence of Productivity: Cross-National Studies and Historical Evidence.* New York: Oxford University Press.

Campbell, Colin, and Jean Laherrére. 1998. The End of Cheap Oil. *Scientific American* 278(3, March): 78–83.

CEA (Council of Economic Advisors). 1998. *Economic Report of the President.* Washington, D.C.: Government Printing Office.

Farrow, Scott. 1985. Testing the Efficiency of Extraction from a Stock Resource. *Journal of Political Economy* 98(3, June): 452–87.

Grossman, Gene M., and Elhanan Helpman. 1994. Endogenous Innovation in the Theory of Economic Growth. *Journal of Economic Perspectives* 8(1, Winter): 23–44.

Halvorsen, Robert, and Tim R. Smith. 1991. A Test of the Theory of Exhaustible Resources. *Quarterly Journal of Economics* 106(1, February): 123–40.

Jones, Charles I. 1997. On the Evolution of the World Income Distribution. *Journal of Economic Perspectives* 11(3, Summer): 19–36.

Kerr, Richard A. 1998. The Next Oil Crisis Looms Large—and Perhaps Close. *Science* 281(21, August): 1128–31.

Lasserre, Pierre, and Pierre Ouellette. 1991. The Measurement of Productivity and Scarcity Rents. *Journal of Econometrics* 48: 287–312.

Lucas, Robert E., Jr. 1990. Why Doesn't Capital Flow from Rich to Poor Countries? *American Economic Review* 80(2, May): 92–96.

Mankiw, N. Gregory, David Romer, and David Weil. 1992. A Contribution to the Empirics of Economic Growth. *Quarterly Journal of Economics* 107(2, May): 407–38.

Meadows, Donella H., Dennis L. Meadows, Joergen Randers, and William H. Behrens III. 1972. *Limits to Growth: A Report for the Club of Rome's Project on the Predicament of Mankind.* New York: Universe Books.

Meadows, Donella, Dennis L. Meadows, and Joergen Randers. 1992. *Beyond the Limits.* Post Mills, Vermont: Chelsea Green.

Miller, Merton, and Charles W. Upton. 1985. A Test of the Hotelling Valuation Principle. *Journal of Political Economy* 93(1, February): 1–25.

Nordhaus, William D. 1992. Lethal Model 2: The Limits to Growth Revisited. *Brookings Paper on Economic Activity* 2: 1–60.

Pritchett, Lant. 1997. Divergence, Big Time. *Journal of Economic Perspectives* 11(3, Summer): 3–18.

Romer, Paul M. 1994. The Origins of Endogenous Growth. *Journal of Economic Perspectives* 8(1, Winter): 3–22.

Schumpeter, Joseph. 1943. *Capitalism, Socialism, and Democracy.* London: Unwin University Press.

Smith, V. Kerry, editor. 1979. *Scarcity and Growth Revisited.* Baltimore: Johns Hopkins University Press for Resources for the Future.

Tietenberg, Tom. 1996. *Environmental and Natural Resource Economics,* fourth edition. New York: Harper Collins.

Tirole, Jean. 1988. *The Theory of Industrial Organization.* Cambridge, Massachusetts: MIT Press.

Young, Denise. 1991. Productivity and Metal Mining: Evidence from Copper-Mining Firms. *Applied Economics* 23(12, December): 1853–60.

2

Innovation and Productivity in U.S. Coal Mining

Joel Darmstadter

This chapter reviews the record of productivity change in U.S. coal mining over the past forty-five years and speculates on the role of technology and other factors in shaping that record. I chose that extended time-span in order to encompass trends prior to, and following, an unsettling ten-year period, beginning in the late 1960s, marked by labor unrest and the impact of landmark health, safety, and environmental legislation. Unlike Chapter 3 of this volume, which discusses the petroleum industry and concentrates principally on productivity in the exploration for and development of new reserves, this chapter concentrates almost exclusively on coal extraction from given reserves, because the overall coal situation in the United States is one marked by vast, economically exploitable reserves—adequate to meet any conceivable volume of domestic and export demand far into the distant future.[1]

In the following pages, a broad review of long-term trends in productivity is followed by a discussion of the key technological developments that have spurred productivity advance in surface and underground mines. That discussion underscores the extent to which investments in advanced equipment and innovative mining practices have complemented the role of labor skills. It thus serves as a useful juncture for introducing evidence on how the more comprehensive (but less readily available) measure of multifactor or total factor productivity tracks the more common measure of labor productivity. The bearing that regulatory policies and labor unrest had on productivity—particularly in the 1970s—is taken up next.

The two subsequent sections look, respectively, at the way in which the U.S. coal industry has structured itself to compete in domestic energy

---

JOEL DARMSTADTER is a senior fellow at Resources for the Future.

markets and as a major contender among world coal exporters. I conclude with brief observations on the problems of and prospects for sustaining coal mine productivity improvement in the years ahead.

OVERVIEW OF PRODUCTIVITY CHANGE: PRE-1970s, 1970s, POST-1970s

The position of the coal industry in America's fuel and power picture remains one of prime importance. Measured in Btu terms, the industry is the country's leading energy producer and an important exporter, its foreign sales yielding annual proceeds of approximately $4 billion. Coal's one-fifth share of the nation's energy consumption is nearly as high as that of natural gas. Coal is the electric utility sector's principal fuel supplier, accounting for around 55% of electricity generated at power stations. Notwithstanding its obligation to meet increasing health, safety, and environmental regulations, the coal industry retains an important competitive edge in the ability to continue serving its traditional markets—especially the electric power sector, whose sustained coal purchases have provided most of the momentum for the industry's viability in recent decades. Absent significantly more restrictive "downstream" constraints—for example, possible limits to greenhouse gas emissions—the Department of Energy's (DOE) Energy Information Administration (EIA) projects coal to retain its ranking importance, amidst stable or declining real prices, into the first several decades of the twenty-first century.

Unquestionably, an important source of the industry's success has been its productivity record and the technological and other factors underlying that record. The broad trends in coal mine productivity, highlighted in Table 2-1, provide a springboard for the detailed analysis in ensuing sections. Underground and surface mining reveal roughly parallel labor productivity trends during the past forty-five years: a record of strong advance in the 1950–60 decade, maintained in the following decade by surface mining, but with some deceleration of growth in underground mines; decisive absolute declines in both sectors during the 1970s; and strong recovery for both since 1980, though it was only midway in this last period that the peak productivity levels attained by underground mining in 1969 and by surface mining in 1975 were again achieved. The nineties have seen some deceleration in the rate of productivity increase, but with a pace that still compares well with that achieved over four and a half decades.

Averaged over that forty-five-year time span, coal mine labor productivity—that is, physical output per miner hour—shows an impressive annual rate of increase of a bit over 4%. Evidently, labor productivity and other efficiency improvements served to contain production costs and permit a concurrent decline in inflation-adjusted coal prices of nearly 1% yearly. The role of those other improvements is captured more explicitly in the estimates presented in Table 2-1, section D. These show that multifactor productivity (that is, productivity based on *all* factor inputs—labor, capital, intermediate goods) tracks the long-term trend recorded by labor productivity alone; however, with nonlabor inputs rising faster than labor inputs, the more comprehensive productivity measure shows predictably much slower rates of increase.[2]

Table 2-1, of course, exhibits only the aggregate and composite manifestations of many interrelated economic, technological, and policy crosscurrents at work during much of this period. Key among these, though in no a priori order of importance, were:

- major technological advances, such as the growth of longwall underground mining and the use of ever-larger excavation equipment in strip mines, in both cases aided by increased computerization and sophisticated control systems;
- the enactment of federal health, safety, and environmental statutes, principally the Coal Mine Health and Safety Act of 1969, the Surface Mining Control and Reclamation Act of 1977, and the Clean Air Act of 1970;
- acute labor unrest, particularly during the 1970s;
- a gradual geographic shift in production toward easily exploitable Western surface coal deposits; and
- a market environment during the oil upheavals of the 1970s in which coal demand—both for current consumption and precautionary inventory buildup—experienced a significant increase. This meant the opening (or re-opening) of small, normally marginal mines and the influx of less-skilled miners, both factors likely to have held back productivity advance.

While unraveling the quantitative effects turns out to be complicated, as we shall see in subsequent sections, it is easy to speculate on the general bearing these and other factors had on coal mine productivity change, positively or negatively, transitionally or more enduringly. Health, safety, and environmental safeguards, irrespective of their arguably positive longer-term benefits and the ability of the industry to adapt to them in due course, could not help but depress productivity levels and growth in the early stages of their implementation. Concurrent conflicts in labor-management relations seemed also to undermine productivity advances. On the other hand, impressive technological developments since the 1970s have brought about strong productivity boosts in both open-pit and underground mines. And within the latter sector, long-

Table 2-1. Productivity Highlights.

A. Average Annual Percent Changes in Labor Productivity (based on short tons/miner day)

	1950–59	*1960–69*	*1970–79*	*1980–95*	*1950–1995*
United States:					
Total	6.5	4.0	–1.9	6.6	4.1
Surface	4.0	4.5	–3.2	6.3	3.3
Underground	6.3	2.6	–3.5	6.0	3.2
Appalachia:					
Surface	3.4	3.3	–5.3	3.8	1.6
Underground	6.3	–2.0	–3.2	6.3	2.3
Interior:					
Surface	5.5	2.9	–3.9	4.3	2.4
Underground	7.4	6.8	–5.0	5.8	4.0
Western:					
Surface	2.3	6.7	–1.3	5.2	3.4
Underground	2.8	1.8	–1.3	8.8	3.6

B. Labor Productivity Levels, 1995 (short tons/miner hour)

		Underground				
			Room-and-Pillar			
	Surface	*Longwall*	*Continuous*	*Conventional*	*Total*	*Total*
United States	8.48	3.85	3.14	2.69	3.39	5.38
Appalachia	3.88	3.39	2.94	2.67	3.08	3.32
Interior	6.39	3.75	3.76	3.67	3.76	4.97
Western	18.93	6.92	4.12	2.60	6.35	15.68

Continued on next page

wall mining, with its comparatively high underground productivity level, increased its share of underground coal production from 27% in 1983 to 47% in 1995.

Amid these developments, one also must note the *aggregate* coal mine productivity implications of geographic shifts. Between 1970 and 1995, Western coal production increased its share of nationwide output from 6 to close to 40%; and since Western coal is dominated by high-productivity surface mining, that shift in itself translates into higher overall coal mine productivity growth. (See Figure 2-1 for more detail on regional production shifts.) Indeed, note from Table 2-1, section A, that, for several of the periods shown, *nationwide* productivity growth *exceeds* the growth rates for the two sectors (underground and surface) comprising the national total. Roughly one percentage point (or one-quarter of the 4.1% rate of productivity increase during 1950–95) is ascribable just to the strength of that westward shift. During the past decade, however, the extent of that shift has slowed perceptibly.

Table 2-1. Productivity Highlights—*Continued.*

C. Percent of Production, 1995

		Underground				
			Room-and-Pillar			
	Surface	*Longwall*	*Continuous*	*Conventional*	*Total*	*Total*
United States	61.6	18.3	16.6	3.3	38.4	100.0
Appalachia	14.8	12.0	12.0	3.2	27.3	42.1
Interior	9.6	2.5	4.2	0.0	6.7	16.3
Western	37.2	3.9	0.4	0.1	4.4	41.6

D. Comparison of Labor and Multifactor Productivity Growth Rates

	Labor Productivity		*Multifactor Productivity*	
Time Period	*Based on short tons*	*Based on Btu*	*Based on short tons*	*Based on Btu*
1970–1980	–1.7	–2.4	–3.5	–4.1
1980–1994	6.5	6.2	3.3	2.8

Notes: Regions are defined as follows. Appalachia includes Alabama, eastern Kentucky, Maryland, Ohio, Pennsylvania, West Virginia, and Tennessee. Interior includes Arkansas, Iowa, Illinois, Indiana, Kansas, western Kentucky, Louisiana, Missouri, Oklahoma, and Texas. Western includes Alaska, Arizona, Colorado, Montana, North Dakota, New Mexico, Utah, Washington, and Wyoming. To construct estimates for Appalachia and Interior for 1950–1980, a rough breakdown was made between east and west Kentucky, which are part of Appalachia and Interior, respectively. All growth rates are compound growth rates. Multifactor productivity growth rates are from Parry 1997.

Sources: Bureau of Mines 1950 to 1976; unpublished EIA data for 1977–1978; EIA 1979 to 1992, *Coal Production*; EIA 1993 to 1995, *Coal Industry Annual.*

TECHNOLOGICAL CHANGE: SURFACE MINING

Emergence of Surface Mining in the U.S. West

In 1950, 75% of U.S. coal production originated in underground mines and 25% in surface mines. By 1974, surface mining had decisively surpassed underground extraction in its share of total U.S. coal production; by 1995, the shares stood at surface, 64%, underground, 36% (EIA 1996a).

By nationwide standards, surface coal—centered on production in Western states, as evident in Figure 2-1—is relatively low in energy content. But even in Btu terms, surface-mined coal still makes up around 56% of the U.S. aggregate. In spite of the relatively low calorific value of Western surface-mined coal, the dual attributes of high productivity and low sulfur content—the latter a significant consideration after passage of the 1970 Clean Air Act—endow the region with a strongly competitive coal industry (Figure 2-2). In 1995, coal prices at the mine averaged $10.15 per short ton in Western states, $18.81 in the Interior coal-mining region,

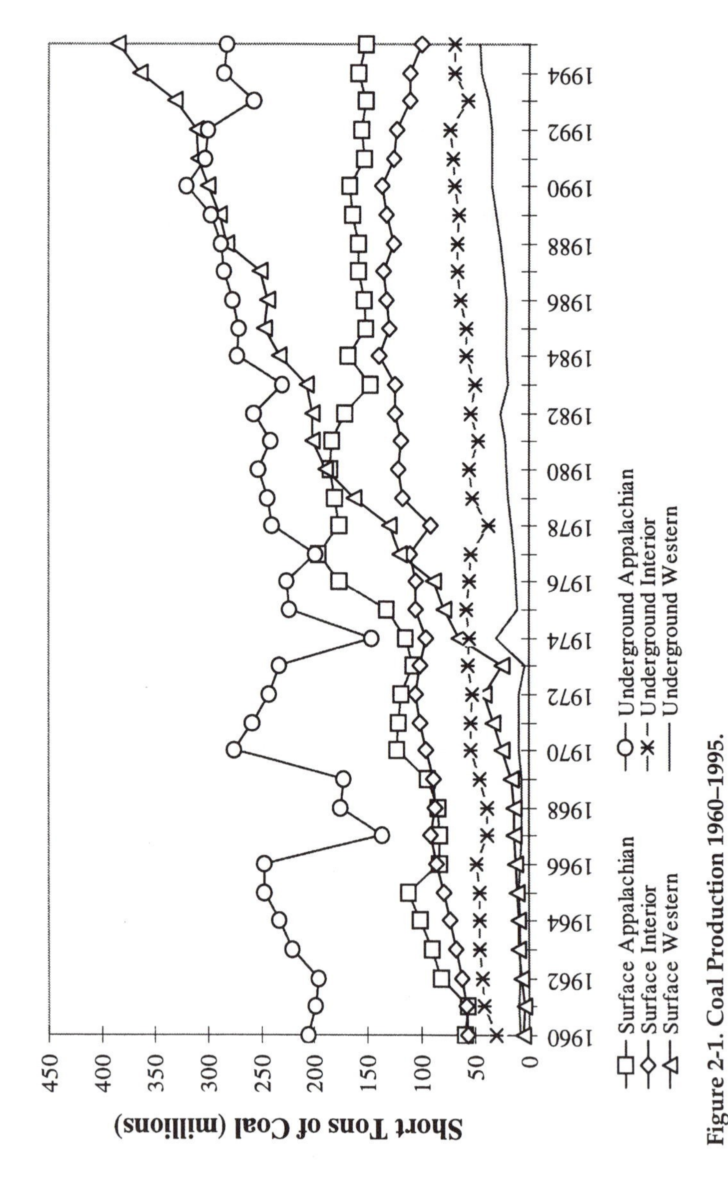

Figure 2-1. Coal Production 1960–1995.

Sources: Bureau of Mines 1950 to 1976, *Mineral Yearbook;* EIA 1979 to 1992, *Coal Production;* EIA 1993 to 1995, *Coal Industry Annual.* Data for 1964 and 1971 are estimated based on previous and subsequent years.

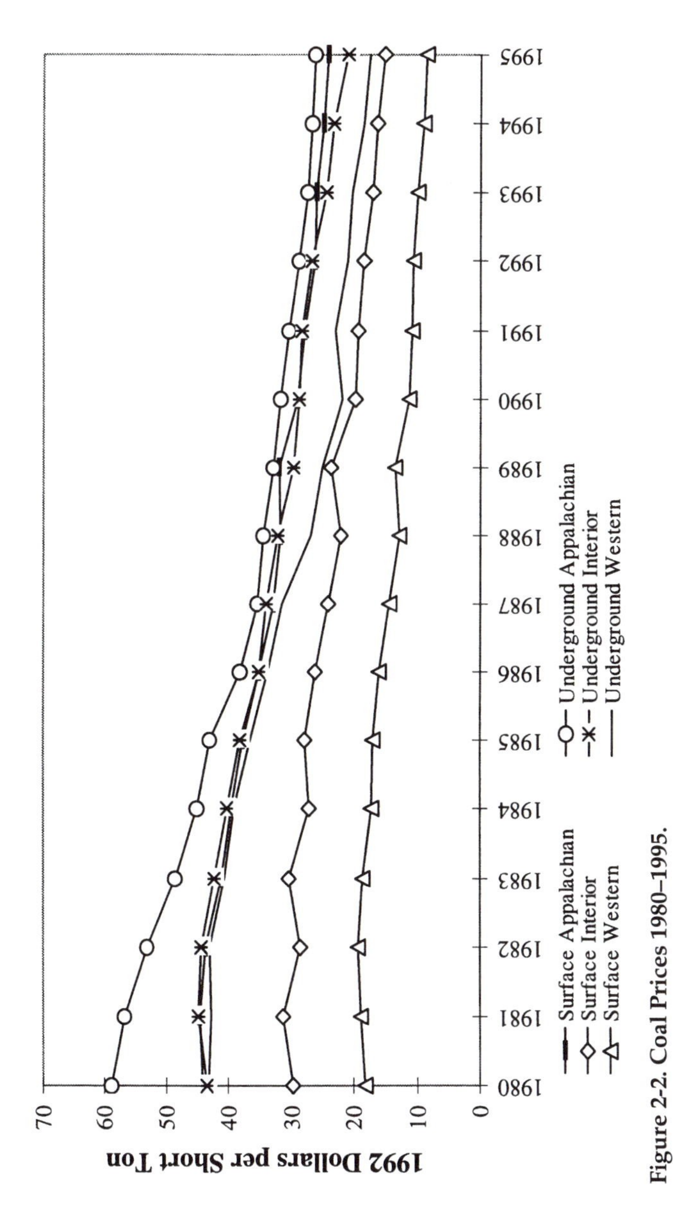

Figure 2-2. Coal Prices 1980–1995.

Sources: EIA 1979 to 1992, *Coal Production*; EIA 1993 to 1995, *Coal Industry Annual*.

and $27.45 in Appalachia. Although Western coal's principal market remains the U.S. West itself, the West's price advantage was decisive enough to allow the region to compete, on a *delivered* price basis, in markets around a thousand miles to the east.[3] For example, in early 1996, some 30% of coal deliveries to Michigan's electric power stations originated in the Western states of Wyoming, Montana, and Utah.

The West's successful penetration of Eastern markets dates from a quarter century ago. In 1971, a mere 4.5 million tons were shipped; by 1976, the volume had risen to 26 million tons (Energy Modeling Forum 1978). What lay behind this rapid expansion? After all, the abundance of the region's strippable coal deposits had been surmised to exist for some time. Several factors now converged to favor exploitation of this potential: the prospects that tight world oil markets would mean increased demand for non-oil energy resources; the increased importance of certain environmental concerns, especially the low-sulfur provisions of the 1970 Clean Air Act; the startup of transport by unit trains, dedicated solely to carrying coal; and, notwithstanding lack of competition in rail transport out of the area, the stability during the 1970s of real average rail freight charges, then still under ICC regulatory control, and the continuing decline of all forms of coal transportation costs in the 1980s and 1990s. (In 1990 dollars, coal rail transportation costs declined from 2.4 cents/ton-mile in 1979 to 1.5 cents/ton-mile in 1993) (EIA 1995a). Concurrently, prolonged disputation over the leasing of federal and Indian coal-bearing lands was veering toward resolution, even if—in some environmentalists' judgment—on terms overly favorable to the coal companies.[4]

Steadily rising output from the prolific Powder River Basin of northeast Wyoming has been facilitated by a concurrent expansion of rail capacity serving the area[5]—both longer track miles and multiple trackage. Distant consuming markets were served, additionally, by mine-mouth generating plants dispatching "coal by wire." A number of such plants arranged to meet their lifetime coal requirements from the adjacent mine. Only a Western surface mine was large enough to provide a power station of several thousand megawatts of capacity with, say, a twenty-five- or thirty-year fuel supply (EIA 1992).

Productivity has without doubt been a central element in causing as well as benefiting from the trend toward surface mining generally and the exploitation of Western open-pit mines in particular. By 1994, the level of labor productivity in Western surface mines—dominated by Wyoming, the nation's largest coal-producing state—was more than 3.5 times as great as nationwide coal mine productivity overall and more than 2.5 times as great as productivity in Western longwall mining—itself impressively high by the national norm (see Table 2-1).

Key Factors in Technological Progress

Aside from poor performance during the period of adapting to significantly changed environmental and related regulations, the predominant experience during most of the last forty-five years has been one of strong productivity advance associated with continued technological improvement—both in the scale and character of the capital inputs to the extraction process and in the skills of the personnel responsible for operating that equipment. An exceptionally striking case in point is Arco's Black Thunder mine in the Powder River Basin of Wyoming, whose annual output of some forty million tons with a cadre of only 500 workers (including those at the coal face and others) dramatizes the marriage of machinery and skills that is the hallmark of a modern surface-mining operation. Nationally, surface mines produce an average of less than 60,000 tons annually, with output per miner about one-quarter the level of Black Thunder. Such a capital-intensive operation also underscores the limitations of just labor (rather than total factor) productivity analysis, as discussed later in this chapter.

In its simplest characterization, surface mining involves the extraction of coal that is exposed once the overburden of earth or rock has been removed. Typically, the minable coalbed lies within several hundred feet of the surface. (Noncommercial deposits may lie at greater depths.) The particular way in which a coal extraction technology is deployed depends on the topography of the mine site—specifically, on whether the location lends itself to *area mining* or *contour mining*. Area mining is used in near-level terrain. Draglines are the dominant technology used to remove the coal and overburden in the area. Once the coal is removed, the overburden is replaced and then the process is repeated until the entire extraction area can no longer be profitably mined. Contour mining is mostly used in mountainous and hilly terrain. The most common variation of contour mining is block-cut mining. In this technique, a box or block cut is made as close as possible to the center of the mining area. The coal is removed from this area; then the overburden from the second area is used to fill the hole from the first area; and so on. Again, draglines dominate the production process.[6]

Differences in extraction technology aside, one can readily identify some of the broad characteristics and general forces underlying surface mine productivity change and levels. Put in the simplest terms, technological advance has been a process marked by progressive improvement and increased scale of existing equipment, augmented by introduction and gradual advances in selected new technologies and practices. Thus, the last several decades have seen evolution of the extraction process from reliance primarily on truck-shovel technology to employment of draglines in conjunction with truck-shovels. At the same time, the capacity of the equipment has been growing, with draglines, for instance, progressing from mobile, diesel-powered units to more powerful "walking" types connected to the electricity grid. In recent years, various phases of surface mine production have also been facilitated by computerization. For example, dragline operations can be positioned and directed with computer-aided analyses of seam thickness and characteristics; global positioning systems have been applied to the truck fleet to improve its performance; and postproduction costs can be contained by computerized systems in coal processing and unit train loading. Increased computing power also has allowed manufacturers to more efficiently design, test, and build equipment. Where this has meant disproportionately reduced capital investment costs per unit of coal output, it is an additional contributor to productivity advance at the mine.

Some Productivity Consequences

Not only have manufacturers improved the overall quality and performance of equipment, but by building equipment to suit specific geological characteristics of seams, they have raised coal recovery. Such improvements have allowed draglines to "fill faster, more easily and more completely" (White 1995). Reporting on the Freedom Lignite Mine in North Dakota, Pippenger (1995) showed how improvements in dragline operations during 1989–1993 raised the volume of dragline production by 18% while dragline costs (adjusted for inflation) declined 23%—in other words, a crudely calculated annual total productivity increase of around 10% in this part of the mine operation.

All told, the complementarity of computerization with advances in the scale and nature of the equipment represents a development of key significance. Information technology and instrumentation—coupled with vast increases in horsepower of equipment—has facilitated the logistical coordination of a variety of stages in the mining operation: cutting, conveying, loading, transport, and a myriad of other activities that occur at a modern surface mine.

The economic benefits of these technological advances, of course, are enhanced when coupled with favorable geologic features—notably, a low stripping ratio of overburden to coal, a characteristic particularly associated with Powder River Basin coal beds. The *stripping ratio* refers to the amount of overburden that needs to be removed in order to gain access to a given amount of coal. Often expressed in physical volume of material per ton of coal, it is sometimes expressed as a ratio comparing the thickness of overburden with that of the coal seam. Intuition would point to the stripping ratio as one important element governing total factor productivity and, thereby, production cost in surface mining. A 1989 nationwide sample survey of thirty-nine surface mines (as reported in the 1992 study by Mutmansky and others) provides statistically significant evidence on the importance of the stripping ratio: the lower its value, the lower average coal production cost. The converse—the risk to profitability of a high stripping ratio—is lent at least anecdotal support by the experience of American Electric Power's Muskingum (Ohio) mine, which—now closed—should, in the judgment of one observer, never have been opened, given its high stripping ratio.[7]

In the overview of productivity change above, we saw that surface-mining labor productivity had advanced strongly throughout the coal-producing regions of the country. The rate of increase was greatest in the West, but significant as well in Appalachia and the Interior region. Whether conditions are favorable for continued nationwide productivity performance of similar or more modest dimensions is problematic. (I revert briefly to this point when I consider productivity prospects and issues below.) Experts have raised questions, in particular, as to whether Eastern coalbed terrain can be brought into production without incurring costlier upfront investments and facing more onerous backend reclamation requirements than in the West. The benefits of scale economies also may be starting to pinch. Experience with ever-larger dragline deployment in mountainous parts of Appalachia appears to have led to some dampening in enthusiasm for significant surface mine expansion. Thus, firms that could start a large-capacity, highly efficient, underground longwall operation would be particularly unlikely to initiate a large dragline/surface mine instead.

TECHNOLOGICAL CHANGE: UNDERGROUND MINING

Some Technological Basics

Among the variety of factors contributing to the strong productivity performance of underground mines in recent decades, one of the most significant has been the emergence of longwall mining technologies. In 1995, longwall mines produced 189 million short tons of coal, about 45% of underground coal production, in contrast to their 20% in 1983. To appreciate the significance of this development, we need to keep in mind that the norm throughout most of the coal industry's history has been the *room-and-pillar* extraction method. Here, the mine roof is supported primarily by pillars of coal dividing the "rooms" which have been cut into the coalbed and where mining takes place. (Figure 2-3 helps bring things to life.) To be sure, even room-and-pillar mining has seen notable techno-

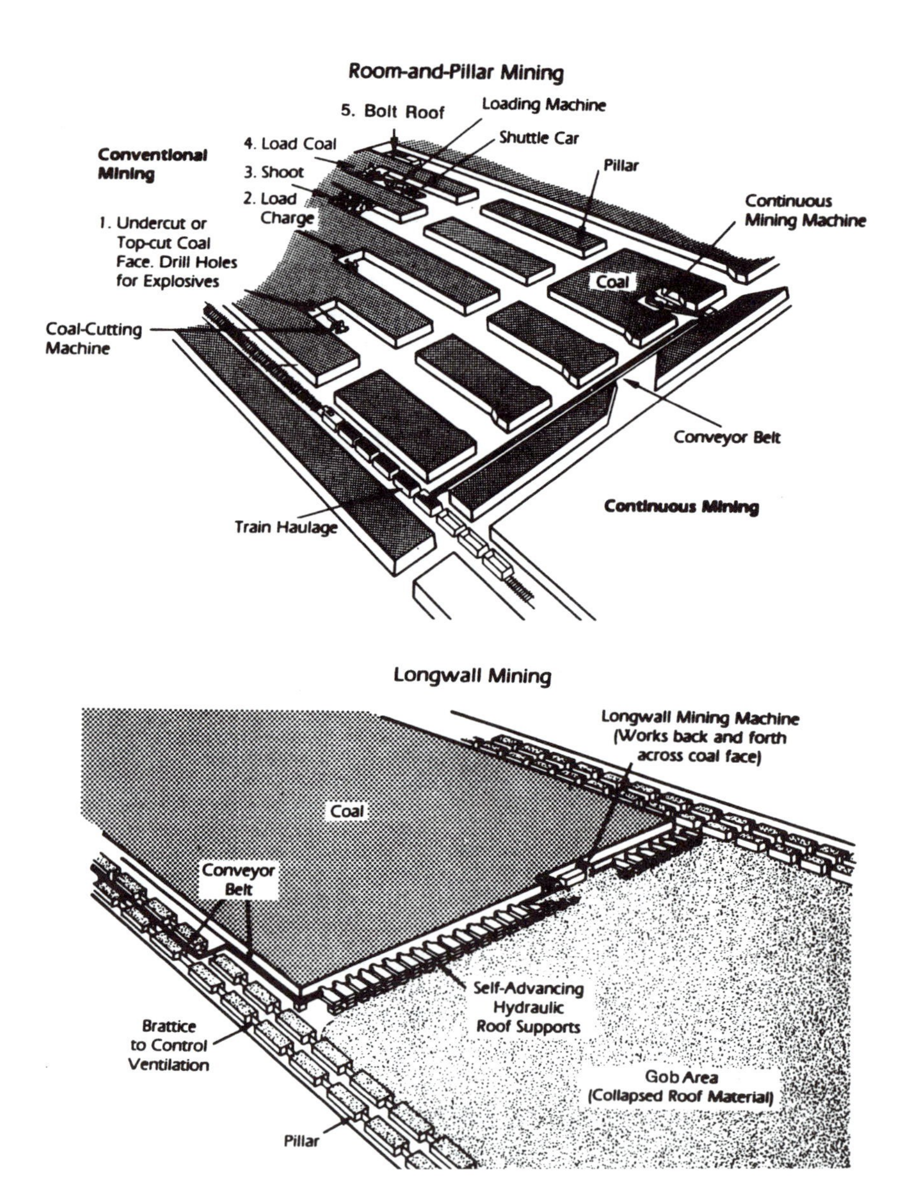

Figure 2-3. Underground Mining Systems.

Source: EIA 1995b, 4.

logical progress; witness the steadily increasing use of *continuous* mining machines, (dating from around 1950), which extract and remove coal from the face in a single operation, in contrast to conventional methods employing a series of separate blasting, removal, and loading operations.

In fact, the arrival of the continuous miner, described by Schurr and Netschert (1960, 312) in their pioneering study as "a technological revolution comparable to the earlier introduction of 'mechanization'," was in its way more of a technological leap forward than the ensuing introduction of longwall extraction. It was the prospect of thinning coal seams that, according to Schurr and Netschert, provided further stimulus for finding ways of ensuring greater coal recovery than seemed possible under room-and-pillar practices.

Although alarm over thinning seams turned out to be premature—the potential of Western surface mining had barely begun to be tapped—development of longwall mining technology was given an important spur forward.

As it turned out, longwall mining, with an average recovery rate of around 57%, succeeds in extracting only a modestly higher proportion of the coal in place than does room-and-pillar mining.[8] However, the typical longwall recovery operation tends to be substantially more cost-effective. (In the following discussion, considerable reliance was put on EIA 1995b.) Longwall mining involves extraction of virtually all the coal contained on a wide rectangular panel, using equipment that allows the roof over the mined-out area of the mine to collapse. (While longwall mining is spared the eventuality of subsidence no more than room-and-pillar mining, subsidence poses fewer problems under longwall conditions, where it is more easily controlled.) The area covered by a longwall mining operation has expanded over the years; by 1993, 82% of longwall units had a width exceeding 600 feet compared to just 12% in 1984. An important factor facilitating that development was the improvement in longwall extraction equipment, especially the approximate doubling in horsepower of both cutting machines and face conveyors. These days, a typical dimension is that of a wall panel measuring some 800 feet wide, with the length of the longwall unit extending about 7,000 feet, all at a height averaging 7 feet. Coal is extracted at a depth—or thickness—of about 3.5 feet. After an initial "blocking out" of a longwall unit (using continuous mining machinery in a room-and-pillar operation),

> Excavation of the coal in the panel is an almost continuous operation. Working under the steel canopies of hydraulic, movable roof supports ("shields"), a coal cutting machine runs back and forth along the 800-foot face, taking a cut ranging anywhere from a few inches to 3-1/2 feet deep during each pass. The cut coal spills

> into an armored chain conveyor running along the entire [width] of the face. This face conveyor dumps the coal onto belt conveyors for transport out of the mine. As the cutting machine passes each roof support, the support is moved closer to the newly cut face to prop up the exposed roof. The roof is allowed to collapse behind the supports as they are advanced towards the face. Mining continues in this manner until the entire panel of coal is removed. (EIA 1995b, vii)

Significant adoption of longwall mining in the United States dates only from the 1950s and 1960s, when the introduction of coal-cutting machines pioneered in Germany served as a spur to American R&D (in part supported by the U.S. Bureau of Mines) and commercialization efforts. (Additional discussion of the international context appears below.) Prior to that time, longwall mining—although understood in principle and employed sparingly—could not compete with room-and-pillar mining; indeed, its labor intensiveness in early applications was the very antithesis of its labor-saving characteristics today.

Productivity

While there are regional exceptions, labor productivity levels in a contemporary longwall operation typically are measurably above those in a room-and-pillar mine. Nationally, longwall productivity recorded a 25% margin of advantage in 1995, compared to parity between the two methods ten years earlier (EIA 1995b, 39–40; EIA 1996c, 82). That advantage is strongly related to the highly mechanized nature of the operation, including opportunities for a significant degree of computerization and substantial continuity in the extraction process. In addition to improving the cutting process, longwall miners, by using a continuously hauling conveyor system, as opposed to a relatively more labor-intensive shuttle car system with continuous miners, are able to increase the rate at which coal is taken to the mine mouth. All these factors combine to make longwall mining surprisingly non-labor-intensive. In a prototypical Eastern mine producing three million tons annually and employing 350 workers, only about 10% of the work force operates at the longwall face itself. CONSOL's Enlow Fork longwall operation in Pennsylvania—the nation's largest underground mine, with longwall dimensions of 1,000 by 10,000 feet—produces eight to nine million tons annually with a work force of some 300 persons. (On an annual output-per-miner basis, this works out to more than four times the national underground average; but, with a diffusion of longwall technology still under way in the industry, Enlow Fork appears to be one of the ranking performers in labor productivity.)

While the growing areal dimensions and application of horsepower in longwall mines went hand in hand with rising productivity, they were not the only factors at work. For example, improved and computerized interaction between roof supports and cutting machines has decreased the number of miscuts that occur and ensures an even and consistent repetition of the longwall cycle. Some experts estimate that avoiding loss of even a few inches per longwall pass can result in savings of up to one day a week (Sanda 1991). Moreover, elimination of the human shield operator decreases risk as well as increasing output. Along with progress at the coal face, parallel advances have increased the efficiency with which coal is transported from the coal face to the mine mouth.

With a doubling of horsepower during 1984–1993 effectively capitalizing on a concurrent increase in face widths and facilitating the tapping of thicker seams in mines of increasingly large size, the proportion of longwall mines producing more than one million tons of coal went from 47% to 70%. These developments provided an important thrust to advances in longwall labor productivity, which between 1983 and 1995 rose from 1.59 to 3.85 short tons per miner hour, an average annual growth rate of 7.6%. Although room-and-pillar productivity trailed that degree of improvement, its own record (5.5% growth yearly) was quite impressive in its own right.

As with other aggregates, this average U.S. longwall vs. room-and-pillar productivity picture varies regionally: longwall enjoys a decisive productivity advantage in the West, in contrast to a much more modest one in Appalachia. EIA conjectures that the disproportionately large share of Appalachian coal destined for metallurgical and export markets involves a sufficiently greater degree of preparation so as to engender some sacrifice in productivity and output.[9]

To be sure, longwall mining presents a number of problems along with its advantages. Size and capital intensity translate into large upfront investment requirements. Productivity during the startup blocking-out phase—a room-and-pillar operation—is typically low. Longwall mining generates substantial amounts of dust and gas that need to be controlled, although longwall mining may not be unique in this regard. In any case, longwall mines generally have better ventilation than room-and-pillar mines. They also have superior safety performance due to reduced personnel at the cutting face. Nevertheless, even in regions where longwall productivity isn't that much superior to room-and-pillar operations, as in the Illinois basin and Appalachia, longwall mines are getting much of the new investment. Evidently, the potential for technologically driven increases in future productivity growth is viewed as highly promising.

In contrast to availability of information about labor productivity in longwall mining, a corresponding record based on total factor productivity (TFP) seems not to exist. Given the enormity of the startup and deferred capital cost requirements, it would be useful to have a sense of what the relatively capital-intensive nature of a longwall mining operation signifies for that more comprehensive productivity measure. A very rough estimate of the ratio of annualized capital to labor costs in the prototypical Eastern longwall mine cited already is about 4-to-3 (calculated from data shown in EIA 1995b, 43–46.)[10] The least that one can say is that, whatever its significance under different economic circumstances, longwall mining has largely demonstrated its economic viability in the context of the low coal price regime that has prevailed in the past fifteen years or so. If only by crude inference, that speaks positively of TFP's role, no less than that of labor productivity, in sustaining the industry's competitive strength over that period.

From the Past to the Future

However striking the productivity-enhancing technological developments in longwall mining over the past several decades, the penetration of a much more pervasive degree of automation appears to offer the basis for further significant productivity improvement in longwall mining (see EIA 1995b, 47, 50). But a number of complementary conditions and factors must come into play for longwall mining to meet such potential. Industry representatives canvassed by EIA elicited certain concerns and pointed to some uncertainties. For example, there may be "...economic limits to the continued expansion of the longwall panel. At some point, the additional capital costs of [greatly extending panel widths beyond the approximately thousand feet that define the industry's current maximum] will exceed the benefits resulting from improved productivity" (EIA 1995b, 57). While firms with large and rich-seamed holdings expect to meet perceived capacity expansion needs via longwalls, experience suggests that producers with small or thin-seamed holdings are likely to revert to room-and-pillar development. In that case, the need for continuous miners with improved cutting rates becomes important. (The lack of such improvement can also hinder the degree of longwall expansion since continuous mining equipment must be used in the development phase of a longwall project.)

There is, finally, the unpredictability of regulatory policy changes that could affect the economics, and therefore productivity, of both longwall and room-and-pillar mining. Such policy uncertainty applies, for instance, to dust control, methane recovery, groundwater integrity, and subsidence protection. As just one example, subsidence is technically inherent in longwall mining (one implication of which is the risk of costly settlements with owners of surface property) but need not be inherent in room-and-pillar mining, provided pillars are left standing rather than mined prior to collapsing. One can see where this could be at least one consideration in determining the economic balance of advantage between one or the other mining strategy.

THE ROLE OF LABOR AND NONLABOR INPUTS IN PRODUCTIVITY CHANGE

The preceding two sections showed that much recent technological change in coal mining has involved application of labor-saving equipment of expanding scale, complemented by increasingly sophisticated computerized processes. Yet, quantification of productivity effects has, of necessity, focused on estimates of *labor* productivity as the most readily available measure spanning extended time periods, multiple geographic regions, and different types of coal extraction. But neither levels nor rates of change in labor productivity capture fully the efficiency with which the entire range of inputs—labor, capital, energy, materials—is deployed in the production process. A comprehensive measure, *total factor productivity* (or, synonymously, *multifactor productivity*), provides a more meaningful guide to the real cost and therefore competitive position of the coal mining industry as a whole or of segments within it. Introduced briefly above and the focus of Parry's crossindustry analysis in Chapter 6, TFP is here accorded a more thorough look. As Parry shows, all of the four natural resource industries surveyed in this volume have exhibited substantial overall similarity in their TFP trends during the past quarter century—namely, sharp rates of decline in the 1970s, followed by strong recovery thereafter. But conditions characterizing coal, petroleum, copper, and forestry were, in some important respects, also unique to each of the four. For a look at the TFP picture in coal mining, most of the discussion in the present chapter is based on preliminary findings in a 1996 draft report prepared by Denny Ellerman and Ernst Berndt (hereafter E-B) of MIT as part of an ongoing research project for EIA (Ellerman and Berndt 1996).[11]

Recognizing that labor productivity data are, as noted, available on a much more disaggregated basis than TFP, E-B set out to test the extent to which labor productivity trends track, or can provide inferences about, the behavior of concurrent trends in TFP. At an industrywide level of comparison, a plot from E-B (Figure 2-4) shows the two productivity paths to generally move together directionally—which is what one would expect—though by no means in parallel fashion. Thus, during both periods of rising productivity shown in the graph (from around 1950 to the early 1960s and from the late 1970s to the present), labor productivity increased at a more rapid rate; while, during the decade of falling produc-

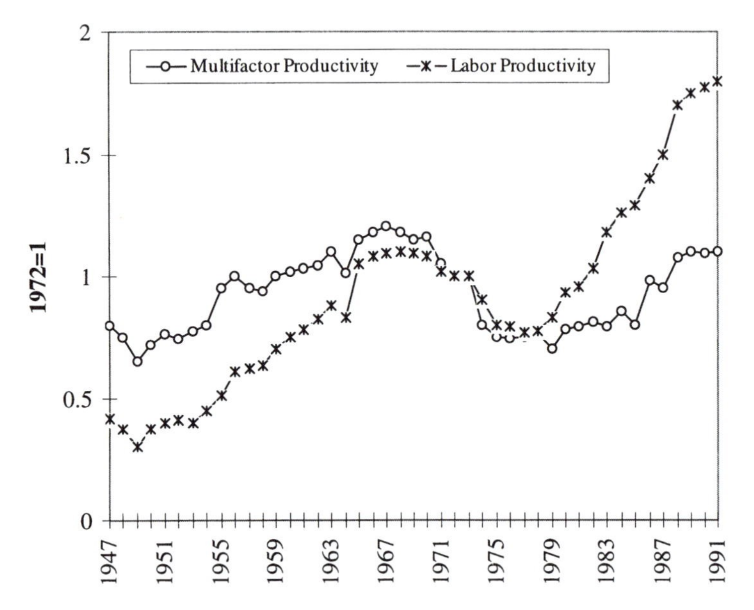

Figure 2-4. Multifactor and Labor Productivity, 1947–1991.

Note: The numerator in the productivity ratio refers to output in tons. The labor component of the denominator has been adjusted for quality change, though that adjustment does not significantly alter a labor input series based on worker hours. See text for additional comments.

Source: Ellerman and Berndt 1996.

tivity in the 1970s, labor productivity decline was less sharp. For reasons that are not clear, preliminary data for the early 1990s show a conspicuous lag in TFP relative to labor productivity growth.

During extended periods of productivity rise, with capital and other resources growing faster than labor, a faster rate of increase in labor productivity than in TFP is almost axiomatic. But in years of productivity decline, as in the decade of the 1970s, the steeper relative fall in TFP arises from the fact that nonlabor inputs—particularly fixed capital, which continues having to be serviced—are not nearly as easy to reduce as miners, who can be laid off.

Conceptually, differences in the rate of change between the two measures of productivity can be ascribed to one or both of the following phenomena:

- *factor substitution:* changes in the shares among inputs—say, more capital relative to labor, but without technological progress—and
- *factor bias:* technological progress associated with one or more inputs, such as the consequences of improved equipment or enhanced labor skills.

Econometric analysis performed by E-B suggests to them that

> factor substitution has not been a major factor in explaining labor productivity in the American coal industry. Most of the observed change of labor productivity is caused by technological change; and the differing rates of change in labor productivity and in total factor productivity reflect the pronounced labor-saving bias of technical change in the coal industry. (Ellerman and Berndt 1996, 13)

Put more prosaically, not only does a worker have more equipment with which to work, but that equipment has become progressively more sophisticated and technologically superior. E-B estimate that "technological progress continually reduces the demand for labor and improves labor productivity at a rate ranging from 2–3% per annum" (Ellerman and Berndt 1996, 12).

Given the heterogeneous nature—regionally and in terms of mining technique—of the American coal industry, E-B proceed to dissect labor productivity trends in eight regions. They do this in order to determine how these disaggregated trends compare with national labor productivity trends and whether they permit conclusions on labor productivity–TFP relationships at the regional scale.[12] This part of the E-B analysis recognizes that the respective impacts on national labor productivity change of compositional shifts, on the one hand, and regional productivity change, on the other, have varied in different periods. But for the extended time span 1972–1994, the shift to high labor-productivity-level regions accounted for as much of the change as the rate of regional labor productivity improvement. However, during the last fifteen years or so, with a marked slowing of the westward shift in coal production, the compositional effect has subsided markedly.

What can be deduced from regional labor productivity trends about regional TFP? E-B believe that regional differences in levels and trends of labor productivity are probably good surrogates for TFP as well. Their chain of reasoning goes as follows. Assuming that the coal industry operates in a competitive environment—both in its input purchases and output sales—coal prices will track production costs that, in turn, reflect what is paid to all factor inputs. Since, as one moves from Appalachian underground mining to Western surface mining, minemouth prices decline roughly proportionately to the increase in labor productivity, an inference can be drawn that TFP regionally may bear the same relationship to regional labor productivity that is exhibited nationally (and depicted in Figure 2-4). To restate it somewhat differently: with enough competition, price is (inversely) proportional to TFP; thus, with price similarly proportional to labor productivity, it must follow that labor productivity tends to be proportional to TFP. But the authors readily admit that their judgments pertaining to the regional scale are quite tentative and invite shoring-up with strengthened research.

At the national level, E-B's findings carry more certainty:

> Although factor substitution is present, the major determinant of the secular trend in labor productivity is change in technology. The pronounced labor-saving bias of technological progress in the American coal industry accounts for most of the observed difference between the rates of change of quality-adjusted labor productivity and total factor productivity.... a reasonable rule of thumb is that total factor productivity is changing at an annual rate 1.5 percentage points less than the rate of change in quality-adjusted labor productivity. (Ellerman and Berndt 1996, 33)

REGULATORY AND LABOR-MANAGEMENT ISSUES: THE PRODUCTIVITY CONNECTION[13]

In reviewing long-term productivity advances (Table 2-1), I commented on the conspicuous break from that trend during the 1970s, when—for every region, type of coal mining, and concept of productivity—the numbers turned strikingly negative. Analysis of what lay behind that break points rather convincingly to the joint effects of health, safety, and environmental regulations to which the industry had to adapt fairly precipitously and, concurrently, to an unsettling and disruptive period of labor-management conflict. But analysis also suggests that, while scarcely trivial—as the following recap of that decade shows—these were largely anomalous and transitory factors in what, on the whole, remains a record of strong underlying momentum in productivity.

Health, Safety, and Environment

Coal mining has long been viewed as an activity posing distinct risks. One associates the health issue prominently with black lung disease or hearing loss; safety with roof cave-ins, subsidence, or methane explosions; and environment with acid mine drainage or unreclaimed spoils from open-pit mining. No doubt, some risks were compensated by wage premiums or company "defensive" outlays to protect against lawsuits, while others were borne by society at large (in which case measured productivity may have been correspondingly overstated). Whatever the case, and notwithstanding pronounced long-term declines in coal-mine fatalities and injuries during preceding decades, landmark legislation introduced in 1969 and 1977 augmented existing legislation with imposition of major new requirements across a broad range of coal mine operations impinging on health, safety, and the environment.[14] The main federal initiatives were the Coal Mine Health and Safety Act (CMHSA) of 1969 and the Surface Mining Control and Reclamation Act (SMCRA) of 1977. (Additionally, and somewhat less directly, the federal Water Pollution Control Act of 1972, subsequently incorporated into the Clean Water Act of 1977, dealt with the impact of coal mining and preparation on water quality; while provisions in the federal Clean Air Act of 1970 helped spur the shift to low-sulfur Western coals.) Among its numerous provisions, the CMHSA addressed such hazards to life and limb as explosive gas mixtures, the integrity of roof support systems, and respirable dust concentrations. Additional legislation, the Federal Mine Safety and Health Act, was passed in 1977. It provided for increased federal mine inspections and created the Mine Safety and Health Administration (MSHA) in the U.S. Department of Labor. Since that time, MSHA has taken over monitoring responsibilities of coal mine health and safety regulations. The two major items covered in the new legislation were coalface illumination standards and "walkaround" provisions, which oblige a worker to accompany a federal mine inspector.

In probing the productivity effects of these new statutes, it is worth appreciating that even costly steps to limit health, safety, and environmental damage can enhance a worker's performance. A miner confident of working under conditions that are not a threat to either health or safety may well be a more productive miner. That said, it isn't hard to see how compliance with the new policies would, at given levels of output, have meant at least a one-time net increase in the level of costs and therefore a penalty in efficiency. For even if compliance did not require additional workers to meet the new standards or require current workers to work in a less "productive" way—thus exacting no penalty in labor productivity—the need for more nonlabor inputs (equipment, materials, energy) would show up as a downward effect on TFP.

What, then, is the empirical evidence on the productivity effects of the new regulatory climate dating from the 1960s and 1970s? What follows are findings from a sampling of three major studies.

- An early effort to probe the issue was a 1981 report of the U.S. General Accounting Office (GAO 1981). Acknowledging the difficulty of disentangling the effects of poor labor-management relations (see below) from those due to the new health-safety-environmental statutes, the GAO found that regulations pursuant to the CMHSA—for instance, compliance with roof control, ventilation and dust control, and various kinds of environmental monitoring—were "a major cause of productivity decline" in underground mines between 1970 and 1973. But, notwithstanding this one-time permanent productivity loss, by the latter 1970s, the regulations were "no longer significant causes of productivity decline" (73). (Nor would one expect them to be causes of further relative changes in productivity.)

- Another early analysis was prepared by Joe G. Baker and his associates at Oak Ridge Associated Universities for the U.S. Departments of Energy and Labor (DOE 1979). Consistent with GAO's findings, this study concludes that the CMHSA is the dominant factor explaining deep mine labor productivity decline from 1970 to 1973, "with its strongest influence occurring in 1973 when ... mine inspections reached an all-time high of more than 70,000. Evidence suggests that after 1973 deep mine labor productivity decline was less related to the CMHSA" (iii). (Negative post-1973 productivity effects appear to have been the result of high market prices bringing lower productivity firms and mines into operation.)
- Edward F. Denison (1985, 66–68), confining his analysis of coal mine productivity to the impact of the 1969 CMHSA, concluded that while the act's productivity growth impacts were pretty well over by 1976, those impacts were substantial for the period 1968–1977. For that time span, as against the Bureau of Labor Statistics' estimate of actual labor productivity in coal mining falling at 3.5% yearly, Denison found that, absent CMHSA, it would have *risen* around 3.1% per year. Underlying that calculation was his estimate that, by 1977, compliance with the act necessitated an increase in the coal mine work force of from 132 thousand to 240 thousand.

This brief survey points pretty much to a consensus view that regulatory policies to protect health, safety, and environmental values had a substantial, if transitory, downward effect on coal mine productivity. Early resumption, after 1980, of impressive productivity growth and technological advance serves to at least attenuate, if not dismiss, the argument by some that the productivity penalty and value of lost output exceeded benefits produced by the regulations of the 1970s.[15]

Labor Relation Impacts

Although labor problems have been a recurrent feature of coal mining's history in the United States, their impact has generally been limited to short-term dips in output rather than basic disruptions with serious spillover effects on the wider economy. The decade of the 1970s was an exception. We concentrate here exclusively on underground mining. For a variety of reasons, surface mining has not been seriously touched by the problem; indeed, that very fact—contributing to the competitiveness of Western coal—might have exacerbated labor problems in Eastern underground mining by increased industry pressure to remain competitive. The strike of 1971 translated into a production drop of 52 million short tons (or 8.5%) below 1970. Later in the decade, a 109-day strike beginning in December 1977 and extending well into 1978, caused an annual decline of nearly 30 million tons—this at a time when the oil-market upheavals beginning in 1973–1974 had substantially increased both domestic and foreign demand for U.S. coal. Indeed, it was precisely labor's determination to share in coal's bull market that helped feed its restiveness. Strikes did not cease with the end of the 1970s, but they greatly abated. In October 1984, for the first time in twenty years, a threatened nationwide strike failed to materialize as the United Mine Workers (UMW) and Bituminous Coal Operators of America agreed on terms for a long-term contract. A new, five-year contract, signed in 1988, averted a walkout as well.

Labor unrest was clearly a significant contributor to productivity decline in the 1970s and probably not just for the two specific years in which the greatest labor strife took place. In probing that question, we need to keep in mind that even without work stoppages, the level of tension, discontent, and sullenness brought about by unending grievances with management probably can't help but condition workplace motivation and efficiency prior to and after a strike, especially one deemed by workers to have been settled on terms inimical to their interest. On the other hand, the effect of strikes per se need not cause a loss in labor productivity. If a closed mine causes a proportionate decline in both production and worker-hours, labor productivity will not be affected, though deployment of overhead personnel and standby maintenance facilities and underutilization of capital that continues to be amortized probably does translate into some productivity penalty, arguably to a greater degree in total factor than in labor productivity.[16] (See the discussion above.)

As for statistical evidence bearing on the labor productivity consequences of labor unrest, GAO saw the labor productivity decline of the 1970s as closely tied to the poor labor-management relations of the period. It particularly singled out the year 1974. In that year, the union contract provision augmenting mine-face equipment operators with helpers is said to explain some 40% of the 13% 1974–1975 labor productivity decline. That is, of course, only one year's record. And the degree of featherbedding implied by the statistics might well be contested by some with a greater hands-on feel for conditions at the mine face. But in another study, surveying that period, Oak Ridge Associated Universities (ORAU) ascribed 25% of the labor productivity lost between 1970 and 1975 to deteriorated labor-management relations[17] (DOE 1979, as cited in GAO 1981, 24).

This review of labor disturbances in the 1970s applies primarily to the unionized component of the underground work force. Although a subject of great interest to academics, government analysts, industry, and, of course, labor, evidence concerning the productivity implications of union-vs.-nonunion labor is ambiguous. ORAU found no statistically significant differences between union and nonunion mines using various productivity measures for the 1970s [ORAU study (DOE 1979) as cited in GAO 1981, 13]. Managers, now as then, take a contrary view, arguing, for example, that nonunion willingness to accept lengthier shifts, quite apart from signifying more *output per shift* (for the arithmetic reason that more hours are being worked) also means greater *output per worker-hour*, since fewer shift changes mean reduced costs associated with the downtime consumed by those changes.[18] In 1995, nonunion nationwide productivity levels in underground mines were, in fact, about 16% above union levels (EIA 1996c). And rates of productivity increase have not significantly differed as between union and nonunion workers in recent years. All told, amidst a rediscovered mutuality of interest between labor and management, labor disturbances have been infrequent, suggesting that the effect of unionization may, in fact, be turning into a minor or nonissue.[19]

INDUSTRY STRUCTURE AND PRODUCTIVITY

Structural changes in the American coal industry over the past several decades have been conducive to rising productivity. The main driving force in that process has been growth in the size both of mines and of firms. Average mine size can rise both because the productive capacity of existing mines is expanded, as in the Powder River Basin, and because newly opened surface and underground mines have tended to be larger than existing ones—further evidence of resource adequacy. Additional contributions to enhanced productivity stem from a number of managerial and work-rule changes.

The relationship of mine size to productivity, illustrated in Table 2-2, arises for several reasons. Large mine size allows for the use of machinery and technologies, such as draglines in surface mining and longwall equipment in underground mining, that would be uneconomical in small-scale operations. Deployment of such large capital inputs is, in turn, facilitated by firms large enough to mobilize the needed investment funds. Increasing the average output of a mine provides opportunities for economies of scale—for example, in more efficiently cutting, processing, and transporting coal to the mine mouth or in economizing in the number of miners needed to perform maintenance and repair work.

Table 2-2. Coal Mine Productivity by Mine Production Range, 1995 (short tons per miner hour).

Mine Production Range (1000 short tons per year)	*Productivity*	
	Underground	*Surface*
10–50	2.03	2.77
50–100	2.36	3.08
100–200	2.72	3.64
200–500	3.45	4.18
500–1000	3.67	4.86
1000 and over	4.11	13.21

Source: EIA 1996c, 85.

The record of growth in the size of mines and firms is a striking one. In 1976, the average coal mine produced 105,000 tons per year; by 1995, output had reached 490,000 tons, with none of the largest fourteen mines producing less than ten million tons per year. Growth in the number of major coal producers—firms with annual output of more than three million tons—paralleled this trend. In 1976, thirty-four firms produced 57% of total coal production. By 1995, the number of major coal producers had reached forty-four, accounting for 80% of total production (EIA 1993; EIA 1996c). Nonetheless, the industry fails to meet criteria commonly invoked to gauge anticompetitive threats. In terms of output shares accounted for by both the top four and top eight producers, the 1995 percentages (33% and 46%, respectively) are well below "market power" trigger points (see White 1987). All in all, coal mining remains a highly competitive industry with a substantial number of firms, relatively easy entry, and—conversely—without the negative consequences for productivity that a more oligopolistic industry, relaxed in its zeal for optimal efficiency, might engender.

In the initial part of the period surveyed here, the industry trend was actually *away* from greater concentration. The trigger for that was the oil-market disruptions and energy price run-ups of the 1970s. Higher oil prices, the expectation of booming demand for coal, and the prospective economic feasibility of converting coal into liquids and gases gave oil companies both the financial wherewithal and the incentive to diversify into coal mining—a move which, at first, involved acquisition of smaller mines and, along with the re-opening of many marginal mines, meant a substantial increase in the number of mines without anything like a commensurate increase in output.

Oil companies weren't the only new entrants into the coal mine industry. In an attempt to ensure a dependable source of energy, many electric utilities started moving toward coal and away from oil- and gas-fired units. Even earlier, a number of these utilities had begun acquiring coal mines or negotiating long-term purchase contracts in order to secure a supply of the desired quality coal for the lifetime of their plants. (Some steel and coke plants followed a similar course of action.) In a detailed historical analysis, Gordon attributed this development "primarily to concern over the logistics of supplying the ever larger plants that the electric power industry was installing. Companies wanted assurance of the availability of the required amounts of coal" (Gordon 1975, 63). Exploiting the transportation efficiency offered by the unit train as an increasingly attractive coal delivery mode reinforced these emerging supply patterns.

The mirror image of the 1970s occurred in the 1980s and 1990s when the combined effect of declining coal prices and recourse to powerful new technologies impelled the departure of many of the small and inefficient mines attracted into the industry a decade earlier. We have seen earlier how, propelled in part by growing constraints on combustion of high-sulfur Appalachian coal and facilitated by the use of ever-larger truck shovels and draglines, the development of large and thick-seamed open-pit Western coal reserves played a major role in increasing average mine size. And similarly to the financial requirements of opening a large surface mine, the adoption and increased use of longwall and continuous mining equipment in underground mining (both East and West) required the assurance of large amounts of output in order to cover the capital costs associated with those technologies

Another change occurred once the price of coal started to decline. Numerous petroleum companies decided to leave the coal business—recognizing the illusion of a viable synthetic fuels industry, selling their coal assets, and, under circumstances where a concurrent fall in world oil prices threatened the profitability of their core business, pursuing cost-reducing technological advances in petroleum exploration and development. (Key developments are summarized in EIA 1993; the petroleum story is told in Chapter 3 of this book.)

The financial impetus for that relinquishment becomes somewhat clearer when one compares profit rates for major energy companies on a consolidated basis with their coal properties only. During the period 1977–1995, while the gap between the two narrowed, returns to the coal component pretty consistently fell below the petroleum components' performance. For the entire period, the Financial Reporting System, a database of thirty-six major energy companies monitored by the EIA, shows the petroleum companies to have achieved an average return of 10% (based on net income relative to net investment in place), the coal firms, one of 4.8% (EIA various years, *Performance Profiles of Major Energy Producers*, Appendix B6). (Some care needs to be taken in interpreting these figures; firms that report to the Financial Reporting System account for more than 50% of domestic petroleum production, but only account for 10 to 15% of domestic coal production.) To be sure, even with its significantly lower returns evidently accepted by asset markets, coal holdings did not axiomatically translate into sell-off properties, even when held by petroleum corporations. The practical explanation may simply be that oil companies came to realize that they were at a *comparative disadvantage* in the coal industry.

Somewhat separate from factors relating to scale, technology, and capital requirements, a number of managerial developments and changes in work practices arguably had positive effects on worker productivity. An example is provided by the large, surface Freedom Lignite Mine in North Dakota. In order to overcome lags in labor productivity advance, the facility in 1992 changed shifts from three eight-hour shifts, five days per week, to two twelve-hour shifts, six days per week. This led to a decrease in the amount of downtime associated with shift changes. With fewer but longer shifts, the amount of time spent moving crews to and from the coal face was decreased, resulting in improved productivity. Shift changes also facilitated more efficient deployment of mobile equipment such as trucks and coal haulers.

Pippenger (1995, 336) cites the case of a company going from a three-shift day, five-day week to a two-shift, ten-hour day, six-day week. The four-hour nonscheduled period each night allowed refueling and preventive maintenance to be scheduled without decreasing the time spent extracting coal. The estimated one-time increase in productivity brought about by shift changes alone ranges from 4% to 12%. To the extent that this example can be generalized to the national level, it suggests that at least some portion of productivity growth—not just in output per worker but in output per *worker-hour*—in recent years can be attributed to changing worker schedules.[20]

Certain other management decisions appear to have strengthened productivity as well. In order to improve the quality of their drilling equipment, some companies have begun signing contracts with equipment suppliers based on a guaranteed number of working hours. This has enabled mines to effectively shift risk away from themselves and onto the equipment supplier, decreasing equipment downtime and increasing available time of equipment at the coal face (Wiebmer 1994). For A.T. Massey, a major Appalachian coal producer, about 50% of all repairs are contracted out in this sort of fashion, the rest (routine maintenance and simple repairs) being performed in house.[21] The earlier practice of most mines retaining their own repair capacity evidently proved economically inefficient. The potential for realizing economies of scale and scope by contracting with specialized firms for major equipment repair and maintenance is part of an outsourcing trend that appears to be growing across a variety of industries. Perhaps, as in coal mining, the increased sophistication and complexity of equipment, instrumentation, and information-system technology are common elements fueling that momentum.

INTERNATIONAL COMPETITIVENESS

With its volume of production exceeded in the world only by China (whose poor-quality coal is consumed almost entirely domestically), the United States, along with Australia, ranks as a major world exporter of both steam and metallurgical coal. Lesser, though still important, exporters in the 1990s include South Africa, Poland, Canada, Colombia, Russia, and Kazakhstan (EIA 1997). Mine-level labor and total factor productivity are not the sole factors entering into international competitiveness; advantages in transportation costs, for example, give Australia the edge in serving Japan's utility market and enable some Canadian exports from the country's western provinces to compete in Pacific rim markets as well. But productivity is clearly a significant determinant of the ability to carve out a commanding share of coal export markets.

Now, since labor skills, along with investments, differ across countries, even a high-wage country can compete internationally if its worker productivity remains high enough to hold unit labor costs, and thereby total costs and prices, in check. That situation prevailed, for example, during the period 1984–1989 when a U.S. Bureau of Mines study found that the U.S. coal mine labor productivity growth rate of 7.4% exceeded growth rates in Australia, Canada, and South Africa. Although Colombia achieved a still higher rate of increase, a surface mine comparison for the late 1980s shows the United States successfully competing with Colombia, a low-wage country whose labor productivity level was estimated at well under half that of the United States (Bureau of Mines 1993). The converse, of course, is that some countries can overcome their low-productivity disadvantage with wage rates low enough to remain competitive.

Also, labor, while a key cost item—averaging around 42% of total production cost in a large U.S. longwall mine—is obviously not the only cost element to be considered. There are capital costs, nonlabor mine operating costs (including various forms of regulatory compliance), land costs, and taxes. Finally, the significance of point-in-time comparisons of world coal trade patterns (and underlying competitive factors to which they contribute) must be validated by trends which can widen, sustain, or shrink the comparative advantage of major coal exporters. For example, the determination by the British and, it appears, German governments finally to eliminate or greatly scale back the subsidies that have long shielded producers from unrestricted foreign competition will give rise to efforts on the part of efficient exporters to exploit the resulting expansion of market opportunities.

The cited Bureau of Mines study, though by now somewhat dated, provides interesting insight into the United States' international competitive standing. Especially striking is the relatively strong showing vis-à-vis Australia, America's major rival (particularly in Asian markets) among world exporters. Not only does the United States show up favorably in labor productivity terms, but that advantage is reinforced by lower labor costs and appears not to be negated by the costlier health-safety-environmental burden then faced by U.S. producers. While more recent data would probably show Australian progress in overcoming the historic fragmentation and strike-prone conditions of the country's craft unions—with accompanying improvements in productivity—labor practices appear still to undermine greater competitiveness with a variety of benefits that raise unit labor costs. For example, Australian coal miners earn an average of 200 hours in annual leave and a maximum of 120 hours in sick leave, compared to 76 and 40 hours, respectively, in the United States. Overall, "lost time"—through such leaves and other absences[22]—constitutes a third of "potential annual work hours" in Australia compared to a fifth in the United States (Walker 1996, 66).

Still, while the U.S. has spearheaded labor productivity advances in recent years, Australia has maintained a sufficiently close pace to remain very much within the front rank of world exporters. Though possessing little more than one-fifth the volume of U.S. reserves, that magnitude and technological sophistication for exploiting its coal resources (involving, in part, direct investment by U.S. firms) suggests that Australia is well-positioned to sustain its important coal trading status.

While South Africa, Colombia, Canada, and several other countries are likely to remain important players in the coal trade, once-prominent coal producers—notably Germany and Britain—are receding from major competitive contention. But it is also a fact—and Britain is a case in point—that mines that survive following closure of uneconomic and inefficient pits sheltered by subsidies compare favorably with technologically advanced operations elsewhere. The volume of production from such modern facilities doesn't add up to dramatic national totals, however.

Prospectively important competitors such as Russia and China have yet to shed the legacy of inefficiency resulting from a distorted pricing and incentive structure under central planning. Even Poland, relatively far along toward economic reform, is still in the process of closing uneconomic mines and transitioning to free markets.

Notwithstanding the retreat of Britain and Germany from major international significance—the inexorable result of unfavorable geological conditions—paradoxically, numerous developments ensuring technologically advanced and safer mining practices among today's leading producers were and, indeed, continue to be innovated by these dominant coal powers of an earlier period. Longwall mining, for example, was beginning to replace room-and-pillar extraction methods in Britain during the early decades of the twentieth century—a period during which it was practiced on only a small scale in the United States. The emergence of longwall mining in Britain and elsewhere in Europe was dictated by gradual increases in working depths to levels where the large pillar requirements of room-and-pillar techniques became increasingly costly (Walker 1996, Chapter 3). Advances in face conveyor equipment developed by Germans after World War II were another important milestone for spurring longwall adoption in the United States. Bucket-wheel excavator technology developed for German lignite extraction is now employed in the Texas lignite industry. It also has been argued that Europe's concern with mine safety—such as methane drainage—preceded and helped intensify such concern in the United States. Additionally, according to the GAO, as of the later 1970s, "European miners are given considerably more safety and skill training in more facets of mining. This increases safety and reduces the disruptive effects of absenteeism. Similar levels of training for U.S. miners may be warranted on safety and productivity grounds" (GAO 1981, vii–viii). Strides during the past several decades in health and safety practices in U.S. coal mining have, of course, been significant. But in assigning credit for such improvements in health and safety as well as in technological progress, the foreign influence deserves to be kept in mind.

That said, the 1996 International Energy Agency (IEA) Coal Research study concluded that, with respect to longwall extraction, all "the evidence clearly indicates that … mines in the United States have gained a significant advantage over comparable operations elsewhere" (Walker 1996, 71). To be sure, U.S. coal achieves some of its comparative advantage from favorable geologic conditions. (In contrast, the progressive deterioration in the quality of U.S. copper ores puts all the more burden on advances in extractive technologies in that industry, as Chapter 4 clearly shows.) But there is more to it than the good fortune of the United States' possessing vast and prolific coal seams. Walker credits the United States with a twofold set of technological achievements: at a general level, impressive techniques of quality control and sophisticated management information systems; at a more specific level, efficient coal handling and conveyor systems capable of maintaining "sustained peak face output without [being] a bottleneck in the process, as has often occurred elsewhere in the past. As a result, longwall mines typically can achieve utilization factors [that is, the proportion of available time that cutting and other machinery is actually operating] of over 60% compared to the maximum of 50–55% attained in Australia, and around 40% in the United Kingdom" (72).

PRODUCTIVITY PROSPECTS AND ISSUES

Aside from observations earlier in this chapter that may have a bearing on future trends, I have not attempted any explicit projections of U.S. coal mine productivity. In this concluding section, I simply cite numbers from a recent EIA energy forecast along with some personal comments. As compared to 1980–1995, coal mine labor productivity to the year 2015 is, as EIA sees it, in for a fairly sharp deceleration in the rate of growth—particularly toward the end of the period and particularly in the case of Eastern, low-sulfur mines facing the need to exploit progressively thinner and deeper coal seams[23] (EIA 1996b, 67–71). Over the next two decades as a whole, EIA projects nationwide coal mine labor productivity to improve at an annual rate of 3.3%. While this is about half the rate achieved during 1980–1995, it is not far off the productivity pace achieved in both Eastern and Western coal over the period 1950–1995. Keep in mind, in this connection, that the more recent period included a decisive shift of coal production from East to West which—given the West's much higher productivity levels—yielded a national rate of productivity improvement exceeding that achieved by either of the two major producing regions. Although some shift from Eastern to Western production is expected to endure, it is projected to be much less pronounced than in the past. In short, judged in a long-term context, rather than the experience since emergence from the dismal 1970s, a 3.3% rate of productivity improvement is not only respectable but healthy enough to sustain level, and even some decline in, real coal prices (see Figure 2-5). Given the prospective rate of improvement in coal mine labor productivity, accompanied by stable or declining coal prices, one can surmise that, implicitly, total factor productivity will continue to rise as well, though, as in the historical case, below rates of labor productivity improvement. Reinforced by a continued decline in its labor demand, inflation-adjusted wage rates in the coal mining industry remain basically flat.

Under these favorable assumptions assumed on the cost side, EIA's "Reference Case" shows real coal prices not only retaining a competitive advantage over oil and gas, but for that advantage to increase with time.[24] Thus coal is projected to retain its dominant share in U.S. electric power capacity and generation. Of course, a decisive price edge is a necessary, but maybe far from a sufficient, condition for coal to enjoy sustained market strength over the next two or three decades. (As noted below, potential environmental constraints could be one obvious barrier to such expansion.) In that respect, EIA foresees a slightly declining share

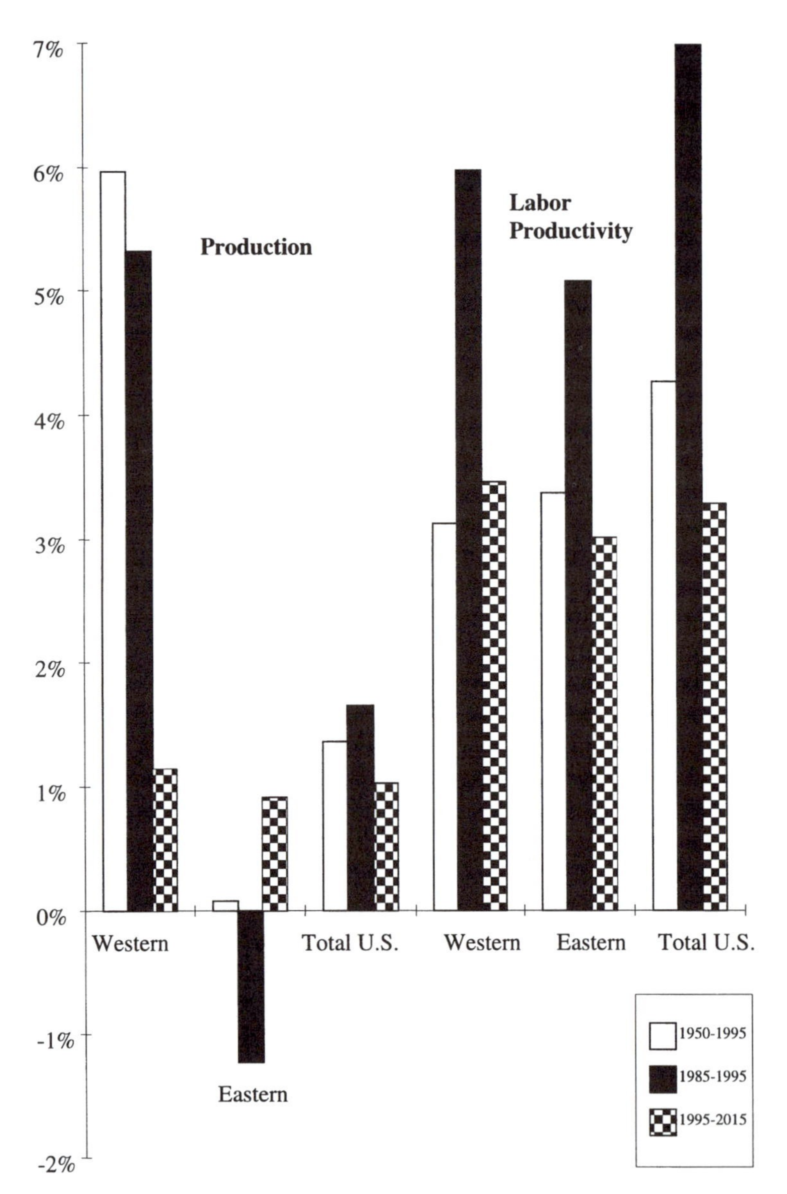

Figure 2-5. Historic and Future Growth Rates for Coal Production and Productivity.

Source: Projection is from EIA 1996b. Historical data from EIA, various years, *Coal Production.*

(though a modest absolute increase) in coal use by power plants and a fairly robust growth in exports.

Although EIA expresses little judgment on the question (perhaps because of its sensitivity about making policy assumptions and judgments), the extent to which technological and other factors that have supported productivity advance in the past will remain equally significant driving forces in the future is a topic that deserves to be at least flagged. For example, might diseconomies of scale emerge, dampening the contribution of longwalls and draglines to productivity growth? Can access to coal-bearing public lands be assumed to be relatively unimpeded? Is the long-term market outlook for coal certain enough to generate the requisite capital investments even if the outlook for such technologies remains promising? Environmental uncertainties may be the biggest factor clouding market prospects and thereby the investment picture.

Even with minemouth coal prices competing favorably with petroleum, pollution control requirements may, on the basis of capital and nonfuel operating cost considerations, justify utilities' opting for gas- rather than coal-fired generation. Such a decision might be contemplated even if environmental constraints arise primarily from local or regional circumstances, for instance, sulfur dioxide or nitrogen oxide emission limits. The implications of a possible greenhouse gas mitigation regime clearly compound the industry's strategic planning dilemma. To be sure, the hesitant pledges for reduced carbon dioxide emissions taken by developed countries at the December 1997 climate change conference in Kyoto, Japan, lacked the certainty of statutory commitment, enforcement rules, and many other implementation provisions, without which the tractability of an effective global warming treaty remains very much problematic. Still, the Kyoto outcome, however tentative, did represent a first step that could within the next few years be followed by more robust measures, with coal combustion being obviously disproportionately impacted as a consequence.

Although health and safety statutes and practices have reduced coal mine injury and fatality rates in recent decades, health and safety conditions will warrant enduring vigilance. In the past, even industry representatives credited the U.S. Bureau of Mines for research that contributed importantly to that improved record. It isn't clear what the Bureau's 1996 demise signifies for future health and safety research. While the final location of the remnants of the Bureau of Mines has not been determined, the expectation is that three of the pre-1996 research centers will be incorporated into the U.S. Geological Survey or MSHA. The largest coal companies have the resources, and probably the motivation, to pick up some of the slack. But, as an "information externality" whose benefits are not easily appropriable and in an industry where there may be only modest potential for intrafirm diversification of risk (as there is for integrated petroleum companies and some other conglomerates), the presumption that such research will become a responsibility adequately borne by the private sector may not be borne out.

It is no doubt premature to judge whether one or more of the uncertainties to which I have alluded is already reflected in observable trends. But several events during 1997–1998 would seem to be at least consistent with a coal-averse, bet-hedging strategy by some industry players, coal producers as well as consumers. A number of major oil companies (Arco, Kerr-McGee, Chevron) are shedding their coal holdings, and some smaller coal companies are being sold off to large companies. In addition, the Edison Electric Institute's power plant construction survey for 1997 indicates that, of seventeen generating units (with a total capacity of some 1,900 megawatts) brought on line, fifteen were natural-gas-fired and none were coal-fired (*Energy Daily* 1998).

I have dwelt here on issues that could prove troubling to the coal industry as it seeks to adapt to the changing market and policy milieu of the first few decades of the twenty-first century. There are no doubt targets of opportunity as well. For example, the need to begin replacing decommissioned nuclear capacity during the next several decades and the potential attractiveness of coal-fired power plants in an increasingly deregulated and intensely competitive electricity market may strengthen coal's long-term market prospects. While my own judgment is that problems are likely to dominate opportunities, it is also true that the last thirty years have seen the industry rise to the challenge of exploiting technological potentials and coming to terms with market realities. And who is to say that it can't happen again?

ACKNOWLEDGMENTS

I have benefited greatly from comments on an earlier draft by Bill Bruno, CONSOL, Inc. (Pittsburgh); Denny Ellerman, Massachusetts Institute of Technology; Hal Gluskoter, U.S. Geological Survey; Richard Gordon, Pennsylvania State University; and my RFF colleagues, David Simpson and Ian Perry. I am grateful to those individuals in industry (both in their offices and production facilities), government, academic institutions, and trade associations who hosted my visits or otherwise responded to queries constructively and with generosity and patience. In that connection, and in addition to those already mentioned, I want to thank Emil Attanasi, David Root, and Tim Rohrbacher, U.S. Geological Survey; Steve Bessinger, CONSOL, Inc. (Morgantown, WV); Hector Choy, James Kliche, and Terry Walsh, Thunder Basin Coal Co./Arco (Wyoming); Eustace Frederick, retired CONSOL official; Andy Gaudielle, Arco Coal Co. (Denver); Willy Hong, Department of Energy, Energy Information Administration; Charles Perkins, Bituminous Coal Operators Association; Stuart Sanderson, Colorado Mining Association; Stanley Suboleski, A.T. Massey Coal Co.; Bruce Watzman, National Mining Association; and Rolf Zimmermann, CONSOL, Inc. (Pittsburgh). (During 1998, Arco Coal's interests were acquired by Arch Coal.)

ENDNOTES

1. In contrast to prevailing annual U.S. production of about one billion short tons, recoverable reserves of coal amount to an estimated 274 billion tons. While technological innovation or higher market prices could raise that estimate still further, environmental constraints—for instance, unacceptability of the high sulfur content that characterizes ninety billion tons of those reserves—could argue for a more conservative, though still very large, figure. The issue is discussed in Darmstadter 1997.

2. The choice of alternative coal mine productivity measures is dictated by the purpose of the analysis and availability of data. In the present study, availability—in terms of disaggregated detail and time-series length—was the overriding basis for the productivity indicator used. Except for selective attention to a total factor productivity measure, our indicator of choice is short tons of coal per hour of labor input. But it is useful to recognize the appropriateness of alternative productivity measures in order to indicate the effect that use of those variants would have on observed long-term trends. Thus, coal output expressed in Btus shows somewhat slower growth than a series expressed in tons, since the westward shift described in the text meant diminishing heat content (but also lower sulfur content). Labor input is most appropriately measured by worker hours, though early years of the historical record are more reliably recorded in number of workers. As noted later in this chapter, statistics on shifts worked are sometimes of interest to those analyzing possible productivity effects of increased or decreased hours per shift. The effect of using several of the output and input series on productivity is plotted in Darmstadter 1997.

3. A recent analysis by Stanley Suboleski shows a 930-mile distance advantage to Western surface mines over Appalachian underground mines, after adjustment for operating costs, royalty payments, and taxes, and assuming equal quality of coal (*Mining Engineering* 1995a; additional information on transport costs appears in *Mining Engineering* 1995b.)

4. For a lively account of the conservationist position, see Toole 1976. The book is particularly telling in its treatment of the revegetation restoration dilemma.

5. In the 1970s, Burlington Northern (now merged with Santa Fe) was the only shipper; presently, the other shipper is Union Pacific (with which Southern Pacific has merged). Additional competition may be in the offing. The Dakota, Minnesota, and Eastern Railroad Corp. has announced plans to try to secure financing for a "$1.2 billion, super-freight hauling line" connecting the coal fields of northeastern Wyoming with the Midwest (*Wall Street Journal* 1997).

6. For more detail on the variety of surface-mine extraction techniques, see Mutmansky and others (1992, 2080).

7. This comment was made by an industry representative at the March 1997 workshop at which a prior version of this chapter was reviewed.

8. For room-and-pillar recovery rates to approximate those in longwall mining requires a practice called *retreat* mining. Here, once coal has been extracted from the rooms, it is removed from the pillars before the roof is allowed to fall.

9. As Richard Gordon pointed out (personal communication), however, even if such a "measured" productivity sacrifice were incurred, the fact that preparation yielded higher *quality* output would signify no necessary loss in "true" productivity.

10. A corresponding room-and-pillar estimate is not available. But my guess is that it would show a markedly lower capital-labor ratio.

11. TFP calculations also receive major attention in Ian Parry's comparative four-industry RFF study (see Parry 1997). To ensure consistency and improve comparability across the four industries (coal, petroleum, copper, forestry), Parry had to employ data sources and estimating procedures that differ to some extent from those employed in the individual monographs and in E-B. For example, E-B rely on mine-level data collected by the Mine Safety and Health Administration, whereas Parry's estimates are based on Bureau of the Census.

12. E-B's procedure involved decomposing changes in national labor productivity so as to separate the effect of changing regional composition of coal production (Powder River Basin surface labor productivity levels, for example, being much higher than Appalachian underground levels) from the effect of regional changes in labor productivity. Additionally, they converted tonnage output figures into Btus in order to adjust for heat-content differentials among regional coals.

13. A more detailed treatment of this topic appears in Darmstadter 1997.

14. Some degree of regulatory oversight over coal mining goes back many decades. The U.S. Bureau of Mines was established in 1910. In 1941, Congress empowered Bureau inspectors to enter mines. And in 1947, Congress authorized formulation of a code of federal regulations for mine safety.

15. One expert who strongly questions the benefit-cost justification for CMHSA is Richard Gordon of Pennsylvania State University. His dissenting judgment was expressed at a workshop discussion of background papers to the present chapter.

16. It is also the case that a more refined measure of TFP would not need to decline, given a suspension of *both* output and the delivery of capital services during periods of curtailed operations.

17. This result was obtained through ORAU's regression analysis utilizing state-by-state variations in shifts lost to wildcat strikes "as a measure of labor discontent" with, in turn, a "strong damaging effect on productivity…" (GAO 1981, 24).

18. In that view, diminishing returns per worker-hour evidently do not set in, or at least, don't have much impact relative to the shift changeover. The same phenomenon apparently exists in the scaled-down British coal industry. Reduction in overmanning and greater shift flexibility have translated into greater productivity.

19. But that mutuality of interest reflects as well weakened union strength: in Appalachia, new mines opened in areas formerly dominated by the UMW are predominantly nonunion.

20. These show a 1986–1995 increase in shift length of from 8.2 to 8.6 hours for underground mines and of from 8.6 to 9.1 hours in surface mining. (EIA 1993 to 1995, *Coal Industry Annual*; EIA 1979 to 1992, *Coal Production.*)

21. Outsourcing of tasks formerly performed by mine personnel could mean some overstatement of labor productivity increases, though not of multifactor productivity increases.

22. Other absences include public holidays, long service leave, and time lost through such things as breaks and shift changes. The "time lost" item is more than 50% greater in Australia than in the United States.

23. The implication that resource constraints (such as rising real costs) may start being felt in these regions illustrates how aggregate recoverable reserve estimates have to be taken with a grain of salt, notwithstanding their definitional characterization as producible at today's prices and technological conditions. See the discussion in Darmstadter 1997.

24. On a Btu basis, the U.S. minemouth coal price stood at 30% of the world oil price in 1995. By 2015, it is projected to fall to 20% (and still no higher than 30%) in the event of a drastic fall in the price of oil.

REFERENCES

Bureau of Mines. 1950 to 1976. *Minerals Yearbook*. U.S. Department of the Interior. Washington, D.C.: GPO.

———. 1993. *A Cost Comparison of Selected Mines from Australia, Colombia, South Africa, and the United States*. Special Publication, August. Washington, D.C.: U.S. Department of the Interior.

Darmstadter, Joel. 1997. Productivity Change in U.S. Coal Mining. RFF Discussion Paper 97-40. Washington, D.C.: Resources for the Future.

Denison, Edward F. 1985. *Trends in American Economic Growth, 1929–1982*. Washington, D.C.: Brookings Institution.

DOE (U.S. Department of Energy). 1979. *Determinants of Coal Mine Productivity Change*. DOE/IR/0056. Study prepared by Oak Ridge Associated Universities for U.S. Departments of Energy and Labor, November. Washington, D.C.: GPO.

EIA (Energy Information Administration, U.S. Department of Energy). Various years. *Performance Profiles of Major Energy Producers*. Washington, D.C.: GPO.

———. 1979 to 1992. *Coal Production*. Washington, D.C.: GPO.

———. 1992. *The U.S. Coal Industry, 1970–1990: Two Decades of Change*. Washington, D.C.: GPO.

———. 1993 to 1995. *Coal Industry Annual*. Washington, D.C.: GPO.

———. 1993. *The Changing Structure of the U.S. Coal Industry: An Update*. Washington, D.C.: GPO.

———. 1995a. *Energy Policy Act. Transportation Rate Study: Interim Report on Coal Transportation*. Washington, D.C.: GPO.

———. 1995b. *Longwall Mining*, DOE/EIA-TR-0588, March. Washington, D.C.: GPO.

———. 1996a. *Annual Energy Review 1995*, DOE/EIA-0384 (95). Washington, D.C.: GPO.

———. 1996b. *Annual Energy Outlook 1997*, December. Washington, D.C.: GPO.

———. 1996c. *Coal Industry Annual 1995*. Washington, D.C.: GPO.

———. 1997. *International Energy Annual 1995*, January. Washington, D.C.: GPO.

Ellerman, A. Denny, and Ernst Berndt. 1996. An Initial Analysis of Productivity Trends in the American Coal Industry. Preliminary draft manuscript, MIT.

Energy Daily. 1998. EEI: Baseload Is Out, Gas Plants, T&D Are In. September 11, 4.

Energy Modeling Forum. 1978. *Coal in Transition: 1980–2000*, Vols. 1-2. Stanford, California: Stanford University, E-17, E-21.

GAO (U.S. General Accounting Office). 1981. *Low Productivity in American Coal Mining: Causes and Cures*. EMD-81-17. Washington, D.C.: GAO.

Gordon, Richard L. 1975. *U.S. Coal and the Electric Power Industry*. Baltimore: Johns Hopkins University Press for Resources for the Future.

Mining Engineering. 1995a. Labor Productivity and Its Impact on Product Price. (July): 658–659.

———. 1995b. Technology Improvements Aid Room-and-Pillar Coal Mining. (December): 1107–1110.

Mutmansky, J.M., and others. 1992. Cost Comparisons. In *SME Mining Engineering Handbook*, 2nd Edition, Vol. 2. Littleton, Colorado: Society for Mining, Metallurgy, and Exploration, Inc.

Parry, Ian W.H. 1997. Productivity Trends in the Natural Resource Industries. RFF Discussion Paper 97-39. Washington, D.C.: Resources for the Future.

Pippenger, Jack. 1995. Competing with the Big Boys: Productivity and Innovation at the Freedom Lignite Mine. *Mining Engineering* (April): 333–45.

Sanda, Arthur P. 1991. Longwall Automation Progresses Slowly. *Coal* (May): 47–49.

Schurr, Sam H., and Bruce C. Netschert. 1960. *Energy in the American Economy, 1850–1975: Its History and Prospects*. Baltimore: Johns Hopkins University Press for Resources for the Future.

Toole, K. Ross. 1976. *The Rape of the Great Plains: Northwest America, Cattle and Coal*. Boston: Atlantic Monthly Press/Little Brown.

Walker, S. 1996. *Comparative Underground Coal Mining Methods*. London: IEA Coal Research.

Wall Street Journal. 1997. Regional Railroad Plans Big Expansion. June 9, A2, A8.

White, Lane. 1995. Advanced Technologies and Loading Shovel Design. *Mining Engineering* (April): 341.

White, Lawrence J. 1987. Antitrust and Merger Policy: A Review and Critique. *Economic Perspectives* (Fall): 16–17.

Wiebmer, John. 1994. Mining Equipment into the 21st Century. *Mining Engineering* (June): 518.

3

Technological Improvement in Petroleum Exploration and Development

Douglas R. Bohi

Exploration for oil and natural gas has changed a great deal since the days when prospects were identified on the basis of surface oil seeps or topographical formations and when drilling was characterized by a group of roughnecks operating a rotary drilling rig and waiting for a gusher to erupt. Over the past two decades, geologists and geophysicists have developed sophisticated seismic techniques to generate mountains of data that are fed into supercomputers via satellites and used to build complex three-dimensional structural models of the earth. Similarly, drillers now use steerable downhole motors to create wellbores that bend and turn at all angles, as well as sensory systems next to the drill bit to determine its location and angle and the composition of the rock layers as they are encountered. And, where not long ago the search for hydrocarbons was restricted to land areas or shallow water, the technology has been developed to explore in water too deep to use fixed-leg platforms. In this environment, remote drilling systems must be used, production platforms must float, and pressures and temperatures are such that oil and gas flowing through subsea pipelines can turn into paraffin and crystals.

Some of the changes in the technology of petroleum exploration and development rival in imagination and expense those involved in exploring outer space. They are a central part of a dramatic story of productivity change that has occurred in the industry in only ten years. This chap-

---

DOUGLAS R. BOHI is vice president at Charles River Associates, Washington, D.C.

ter describes the changes in technology and their contribution to lowering costs and boosting productivity in petroleum exploration and development.

The term productivity is used in the conventional way to mean the amount of output that can be produced with a given amount of inputs. However, the measures of outputs and inputs used in the context of petroleum exploration and development are not conventional. For example, for the measure of output it is more meaningful to refer to the level of success in finding new discoveries rather than the amount of oil and gas produced. Similarly, the measure of inputs refers to the number of wells drilled or seismic crews at work rather than the more conventional number of labor hours or amount of capital investment. More will be said about the measures of productivity below.

The motivation for the changes in productivity described here results in part from increasing pressure to reduce costs of production. The decline in the prices of oil and natural gas in the 1980s and the prospect that prices would not rebound in the foreseeable future meant that firms had to find new ways to reduce costs in order to turn profit margins around. The story of productivity change in the industry is very much a story of how the industry responded to the pressure to reduce costs. The requirement to reduce costs had to be achieved, moreover, in spite of the inexorable effect of depletion working in the opposite direction.[1] As reserves are found and extracted over time, it becomes more expensive to replace them. Unless technology improves to make it easier to find and produce the next generation of resources, the average cost of production will inevitably rise over time. Thus, the pressure to lower costs is closely linked to the need to develop new technologies.

The next section describes the overall performance of the U.S. petroleum industry in recent years to give a perspective on the magnitude of the changes taking place. The following three sections discuss the implications of the three most important technological innovations to come along in the last ten years: three-dimensional (3D) seismology, horizontal drilling, and deepwater production systems. The final section discusses the origins of these technologies and some implications for the competitiveness of the U.S. industry and for the world price of oil.

EXPLORATION AND DEVELOPMENT ACTIVITY IN THE UNITED STATES

Exploration and development refer to all of the upstream activities that are required to find new oil and natural gas reserves and develop them to the point where they are ready to be produced. Expenditures on exploration and development in recent years have been running at a level of $10–$15 billion per year, or about a third of what they were in the early 1980s. The explanation for the decline is simple. The incentive to invest in exploration and development is determined by expected earnings, which is determined in turn by the market prices of oil and natural gas. While investment and revenues have declined, the industry is not neglecting upstream activities. The ratio of exploration and development expenditures to net oil and natural gas revenues has been increasing since 1987. This ratio, sometimes called the *plowback ratio,* indicates that the industry increasingly has been willing to commit its financial resources to invest in new reserves, even while the prices of oil and natural gas were falling.

What the industry has been buying for its exploration and development dollars also has changed in recent years. There has been a shift from onshore to offshore projects, particularly toward costly, but profitable, deepwater projects in the Gulf of Mexico. Another change is the target of drilling efforts: from oil to natural gas. By 1993, the number of successful gas wells drilled exceeded that of oil wells for the first time in U.S. history.

To measure how productivity in exploration and development has changed, we shall look behind the usual measures that focus on the ratio of oil and gas produced to the number of workers employed (or to the total amount of factor inputs).[2] Such measures provide an inaccurate picture of how productivity has been changing in recent years because firms may alter the rate of extraction from available reserves in response to, say, a change in the price, rather than as a result of a change in the inventory of reserves available for production. Another consideration is that it may take some time before a change in upstream productivity results in a change in output. For example, the new technologies under consideration here may lead to an immediate improvement in drilling success rates at the margin, but will have a delayed effect on the average success rate, and an even longer delayed effect on the level of output.

A better picture of productivity change in exploration and development is obtained by looking at the direct results of upstream activities involved in adding new reserves, such as drilling success rates, discovery rates, discovery sizes, and finding costs. While no one of these measures captures all aspects of upstream performance, together they can provide a good picture of how productivity has changed. As we shall see, all of them indicate that there has been a significant improvement in productivity in recent years.

Figure 3-1 shows the overall success rates for exploratory and development wells, including both oil and gas wells, and shows a dramatic improvement in exploratory success after 1990. A success rate in this context is the percentage of exploration and development wells drilled that

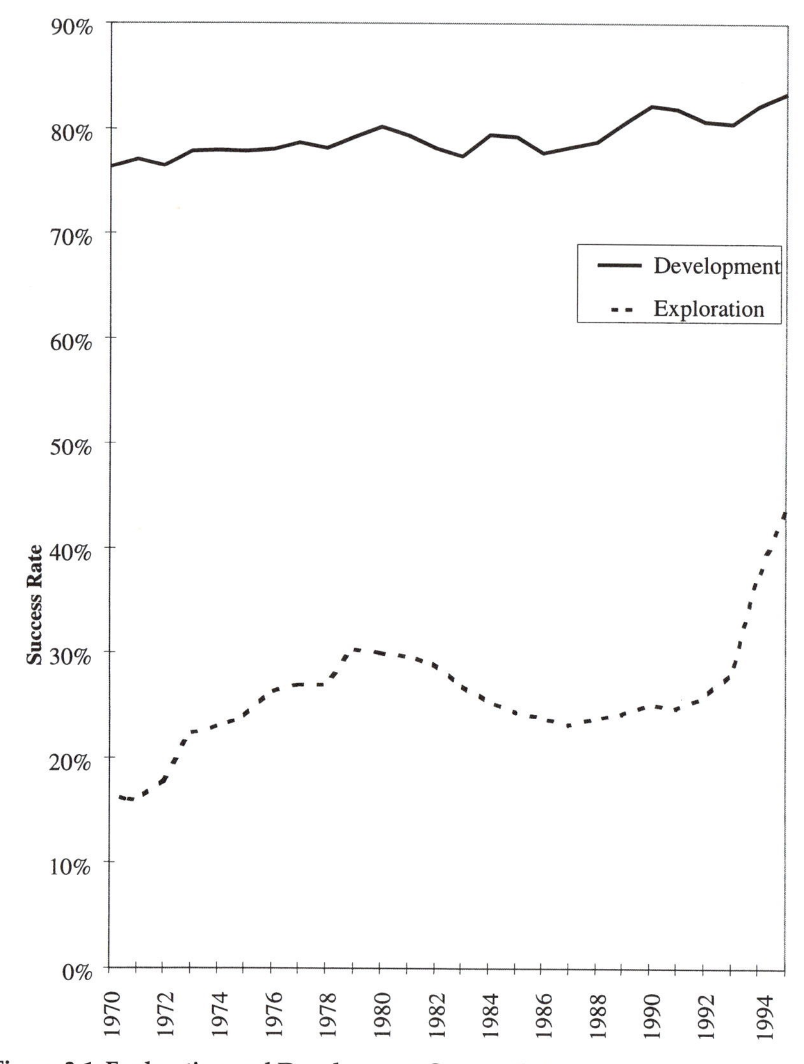

Figure 3-1. Exploration and Development Success Rates.

Source: Energy Information Administration 1996.

hit a commercial deposit, regardless of the size of the deposit or the potential extraction rate. Since development wells are drilled into known reservoirs, while exploratory wells are intended to find new deposits, the average success rate of development wells is understandably much higher than that for exploratory wells.

The same dramatic increase shows up in the rate of new field discoveries per new field wildcat well drilled (see Figure 3-2). The *wildcat* defini-

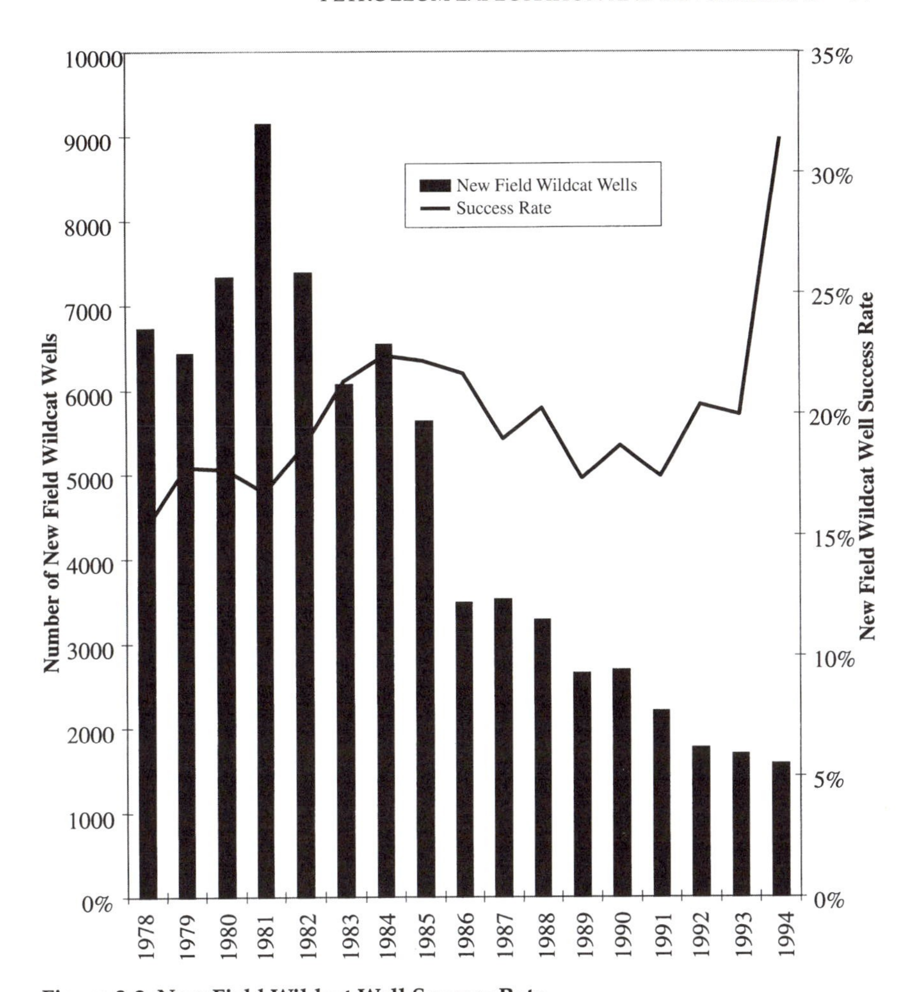

Figure 3-2. New Field Wildcat Well Success Rate.

Sources: Energy Information Administration 1996; American Petroleum Institute 1996.

tion of success rate refers to the discovery of new resources in virgin territory, in contrast to the *exploratory* success rate, which includes discoveries in established as well as new areas. As expected, the success rate is somewhat lower for wildcat drilling than for exploratory drilling, but they both show the same sharp improvement after 1990.

The average amount of oil and natural gas contained in each new discovery is shown in Figure 3-3. Average discovery size for both oil and natural gas was trending downward until 1986 but has increased in recent years. The turnaround in average discovery size in recent years is

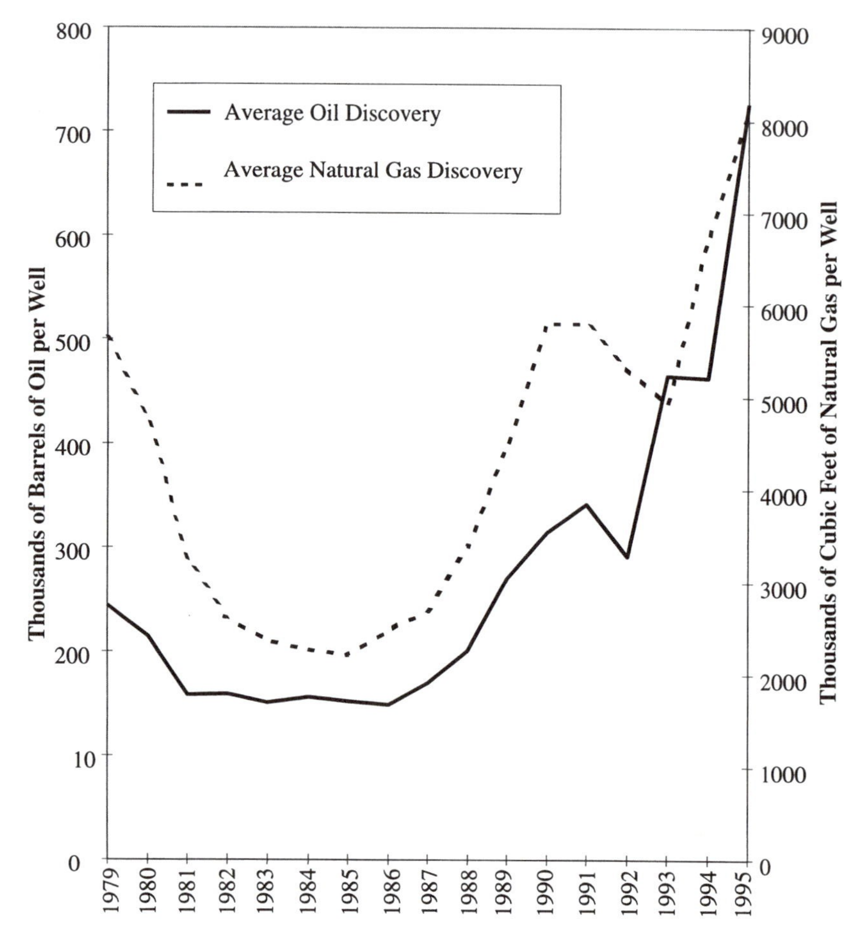

Figure 3-3. Average Discovery Size (three-year moving average).

Note: The quantity of the hydrocarbon discovery is defined as the sum of discoveries in new fields, discoveries in old fields, and extensions.

Sources: Energy Information Administration 1996; Energy Information Administration, various years, *U.S. Crude Oil, Natural Gas and Natural Gas Liquid Reserves Annual Report.*

especially surprising for two reasons. First, discovery size is continuously updated over time, as continued development of reservoirs leads to extensions and revisions of the amount of resources in-place that are credited back to the original discovery. Thus, more recent figures on average size are biased downward relative to older figures simply because there has been less opportunity for revision. Second, discovery size should fall over time as the resource base becomes more depleted and the larger fields are discovered first. A reasonable hypothesis is that the new technologies described below have been effective in finding and exploiting larger deposits.

For comparison with the above measures of performance, "yield per unit of effort" measures additions to oil and gas reserves per dollar spent on exploration and development activities. Exploration and development expenditures are commonly divided into two separate measures of the "level of effort" to reflect the distinct activities involved. Exploration expenditures are aimed at finding new discoveries, but this measure is ambiguous because the amount that is discovered in one year changes over time with revisions and extensions (Adelman 1997). Until a reservoir is fully developed, the magnitude of the initial discovery and the yield per unit of effort are unknown. The outputs of development expenditures, in contrast, are revisions and extensions, which are known for each year in which development drilling takes place.[3] Whether measured on the basis of exploration or development activity, however, the yield per unit of effort for both oil and gas drilling (Figure 3-4) has increased steadily since the early 1980s and especially after 1990.

The inverse of "yield per unit of effort" is the "average finding cost" of new reserves; that is, expenditures on exploration and development divided by additions to reserves. As shown in Figure 3-5, the average finding costs for oil and natural gas have both declined by about 50% since 1982. It should be emphasized that the level of average finding cost is less important than the trend over time. The downtrend summarizes the dramatic improvement in productivity in petroleum exploration and development in the United States.

3D SEISMIC TECHNOLOGY

This is the first of three sections that discuss the implications of recent advances in exploration and development technologies. Three-dimensional (3D) seismology is by far the most important of the three, in large part because of its remarkable success in finding and developing reserves, but also because of its contribution to the success of the other two technologies.

A Description of 3D Seismology

Reflection seismology uses sound waves propagated into the earth and reflected back to the surface to infer the structure and properties of subsurface rock layers.[4] The technique has been used in the search for hydrocarbons since the 1920s, but a revolution has occurred since 1985 with the development of 3D seismic methods. Compared to the earlier two-

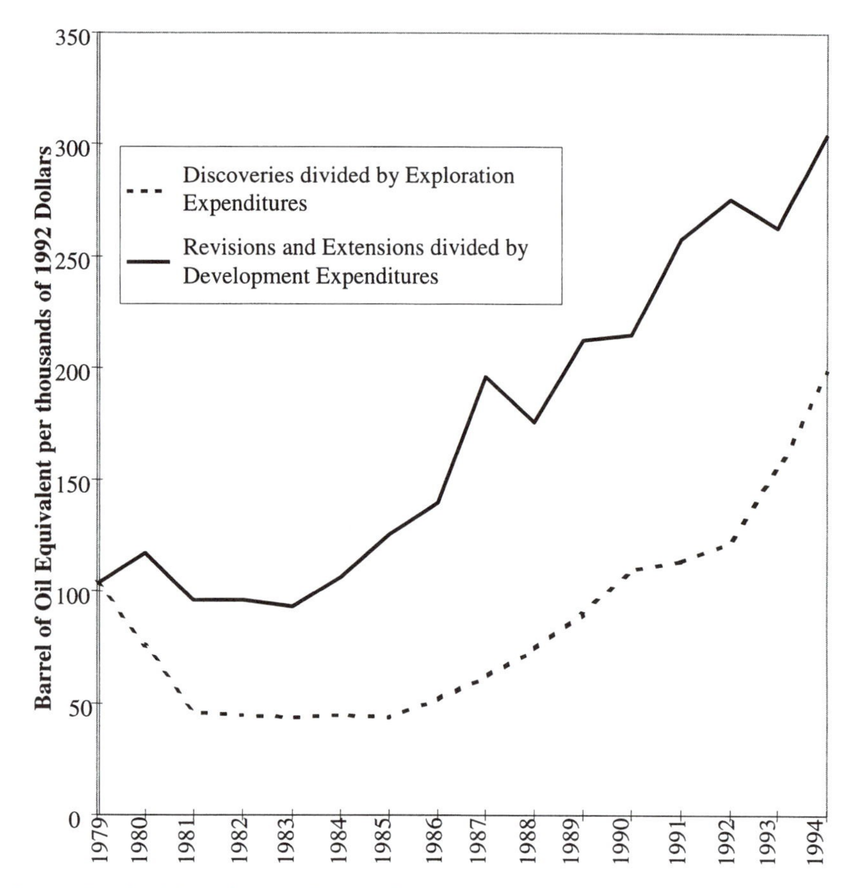

Figure 3-4. Yield per Unit of Effort (three-year moving average).

Note: Barrel of Oil Equivalent is defined as a weighted sum, based on Btus of oil and gas discoveries, revisions, and extensions.

Sources: Energy Information Administration 1996; Energy Information Administration, various years, *U.S. Crude Oil, Natural Gas and Natural Gas Liquid Reserves Annual Report.*

dimensional (2D) methods, 3D seismology provides a better picture of the composition and structure of subsurface rock layers. The extent of the improvement has been compared to the improvement in magnetic resonance imaging over X-rays in medicine. The higher-quality images greatly improve the ability to locate new hydrocarbon deposits, determine the characteristics of reservoirs for optimal development, and determine the best approach for producing a reservoir.[5]

Seismic surveyors generate sound waves at specific locations on the earth's surface or in water that travel into the earth. Part of the energy is reflected back to the surface (hence, "reflection" seismology) whenever it

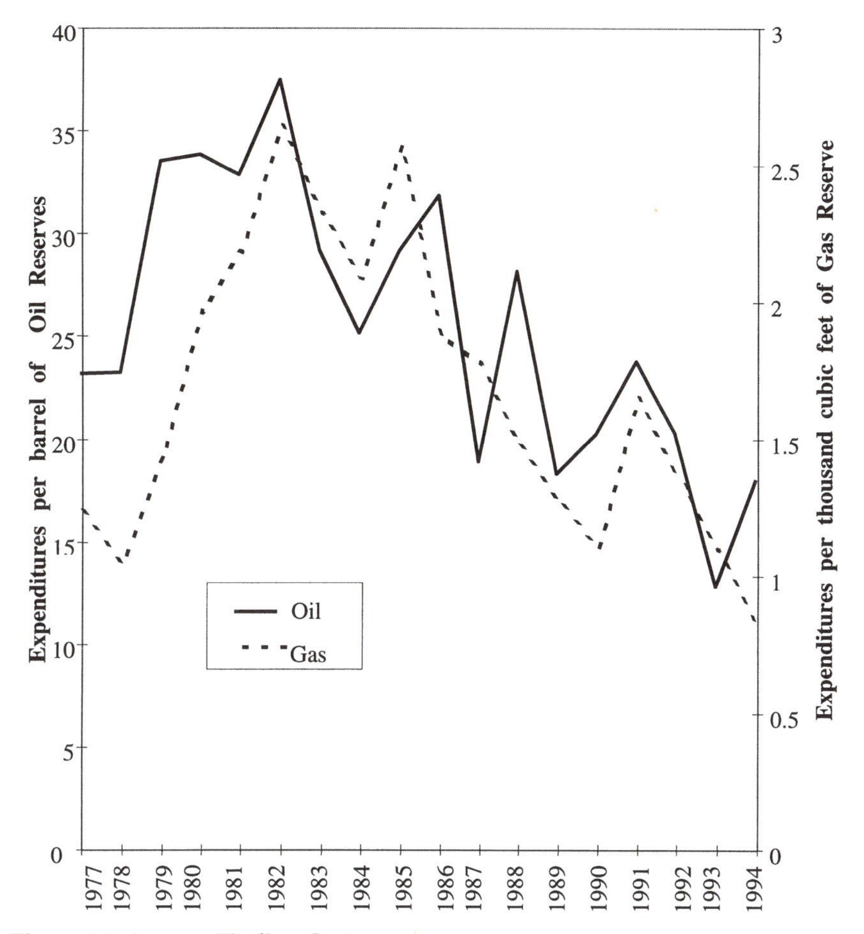

Figure 3-5. Average Finding Cost.

Note: Average finding cost is the ratio of exploration and development expenditures to the sum of discoveries, extensions, and revisions.

Sources: Energy Information Administration 1996; Energy Information Administration, various years, *U.S. Crude Oil, Natural Gas and Natural Gas Liquid Reserves Annual Report.*

passes through a discontinuity such as the interface between two rock layers of different densities. The energy is reflected back to receivers on the surface called *geophones* (on land) or *hydrophones* (in water), where the reflected waves are converted to electrical signals that can be recorded for subsequent analysis. The receivers measure the arrival time of the sound waves to determine the location of subsurface structures and the amplitude and frequency of sound waves to determine the composition of rock layers.

An enormous amount of data is required to yield a high-resolution image of the subsurface, and 3D methods require much more data than 2D methods. In fact, the lack of high-speed computing capability delayed the emergence of 3D technology. Once parallel computers arrived in 1985 to solve the processing problem, it did not take long to develop improved sound receivers, data handling and transmission equipment, and analytical models to take advantage of the innovation. Additional time was required to gain practical experience with the use of 3D seismic information from successes and failures in the field. Anyone could purchase 3D data and process the data using dedicated computer workstations, but interpreting the results for oil and gas potential is a combination of art, science, and experience.

The difference between 2D and 3D methods is related to the way data are generated. A 2D survey consists of a sound source and a series of receivers laid out along a single line (referred to as an azimuth) at specific distances from the source.[6] The source emits a sound wave that travels through the earth and reflects off the impedance contrast between layers.[7] Computers calculate the paths of the reflected signals and the depth of the rock layers. With many receivers laid out on a line from the sound source, many such depth calculations can be made. The sources and receivers are then moved, along the same azimuth or to a parallel azimuth, to map a larger subsurface area. The continuous movement of sources and receivers creates overlapping, or redundant, samples of data for the same subsurface location. The redundancy of observations raises the signal-to-noise ratio and improves the resolution of the subsurface image.

To obtain a high-quality resolution of a specific subsurface area requires seismic shooting over a much wider surface area because many redundant shots of the same subsurface points are required using source positions at different, and sometimes distant, points on the surface. The area of high-quality resolution looks like the small end of a cone that extends from the subsurface location to be mapped up to the wide end at the surface. The wide end of the cone at the surface indicates the area that must be surveyed in order to obtain a high-quality image for the small area below. The further one moves away from the cone, the fewer redundant observations are received for the same subsurface location and the lower the quality of resolution at that location. Also, the deeper one goes in mapping a particular subsurface structure, the larger the area of seismic shooting on the surface required to image that subsurface structure, and the higher the cost of the survey.

In contrast to a 2D survey, which collects data along a given azimuth of the earth's surface to interpret a vertical cross-section of the earth beneath the azimuth, a 3D survey positions receivers with respect to the sound source so that each shot will be received along a variety of different azimuths. This variation in azimuths is an important element that accounts for a dramatic improvement in the resolution of 3D surveys relative to 2D.[8] The large number of observations associated with each shot also accounts for the enormous increase in data from 3D surveys relative to 2D surveys.

Using 3D data, a three-dimensional model of a subsurface area is constructed that may be dissected either horizontally or vertically. A vertical slice gives the same picture of a cross-section of the subsurface geologic structure as 2D data, only better. Horizontal sections, also called time slices because they represent different time periods in which sediments were deposited, can be literally peeled off a few feet at a time to reveal ancient river channels and sandbars. Such a planar view would be impossible without 3D seismology. This information is valuable in understanding the processes by which sedimentary layers were formed.

One indicator of the significance of 3D technology is that it is now possible for the first time to map geologic structures below salt layers. Seismic waves pass through salt much faster than adjacent sediments, causing the base of the salt structure and deeper seismic reflections to appear higher than they really are. In addition, the sound waves are refracted in unpredictable ways so that noise rather than signals are recorded.[9] 3D techniques increase the chance of correctly positioning the location of subsalt structures and of mapping the interfaces between salt and sediment layers. One indication of the improvements achieved with 3D methods is the first discovery of a subsalt hydrocarbon deposit by an Exxon–Conoco team in 1990 and the first discovery of a subsalt commercial deposit by a Phillips–Anadarko–Amoco group in 1993, both in the Gulf of Mexico.[10]

Ironically, 3D seismic surveying was not used initially to explore for new hydrocarbon reservoirs, but to aid in the development of known reservoirs. The reason is that in the early years, the cost of a 3D survey was at least three times that of a 2D survey, so applications of 3D were directed to where it was thought the investment had the greatest payoff. A 3D survey could be restricted to the immediate vicinity of a known reservoir, while 2D was used to survey broad areas to find new reservoirs. 3D immediately proved its worth in development applications because it is capable of revealing minute details about a reservoir that are not possible with other methods.

In addition to its role in exploration and development, 3D seismic information is used in monitoring a producing reservoir over time. Adding the time dimension by acquiring a time-sequence of 3D images, called 4D seismology, enables producers to manage the depletion of an oil reservoir more efficiently than before by monitoring the penetration of water or gas into the reservoir and avoiding "coning" around the well-bore.[11] The same procedure would be followed to monitor the progress of enhanced oil recovery methods such as water flooding, where water is pumped into the reservoir through injection wells in order to force oil through the reservoir and toward producing wells.

Implications for Productivity

There are no comprehensive data available that may be used to calculate the precise productivity implications of 3D seismology. What is available are the anecdotal experiences of individual companies that have published reports on their drilling activities that compare their experiences with the use of 2D and 3D seismic techniques. Combining these experiences provides a good picture of some of the benefits of using 3D seismology, at least in terms of the successes and failures in finding and developing reserves.[12]

Table 3-1 summarizes the experiences of five companies regarding their comparative success rates using 2D and 3D seismology for exploration and development drilling in different locations and widely different circumstances. Because of the differences, the numbers cannot be given equal weight in summarizing the benefits of 3D. Based on a subjective evaluation of the estimates, as indicated below, it is concluded that 3D is responsible for increasing the exploratory success rate from about 20% to about 50%, and the development success rate from about 70% to about 85%.

In reaching this conclusion, the greatest weight is given to the Amoco experience reported in Table 3-1. Amoco created a 3D Seismic Network of Excellence group to, among other things, assess the value of using 3D seismic methods within the company. Reports published by Aylor (1995a, 1995b, 1996) as a result of this effort represent the most accurate, comprehensive, and detailed look at the advantages of using 3D. The results are based on data collected for 159 3D surveys conducted during 1991–1994 in several areas of the world.

The surveys are divided into two groups, labeled exploration and exploitation, where the former refers to wildcat exploration and the latter includes new field exploration and older field extension. The numbers that appear under the label "development" in Table 3-1 come from the exploitation group. The results are based on 998 production wells that were drilled with and without the benefit of 3D. The experience with exploration wells refers to a total of 272 wells, of which 38 were based on 3D information and 234 were not. As indicated by Aylor (1995b, 77), these numbers reflect the fact that "Amoco in the past has used 3D quite late in field development." Like many companies, Amoco at first used 3D pri-

Table 3-1. Drilling Success Rates: With and Without 3D Seismic Technology.

Company and Location	*Exploration Drilling*	*Development Drilling*
Amoco: Worldwide		
With 3D	.48	.86
Without 3D	.13	.57
Exxon: Gulf of Mexico		
With 3D	.70	
Without 3D	.43	
Exxon: United Kingdom, North Sea		
With 3D	.47	
Without 3D	.36	
Exxon: Netherlands Offshore		
With 3D	.70	
Without 3D	.47	
Fairfield Industries: Gulf of Mexico		
With 3D	.50	
Without 3D	NA	
Fairfield Industries: Louisiana Shallow Water		
With 3D	.57	
Without 3D	.20	
Fairfield Industries: Louisiana Onshore		
With 3D	.43	
Without 3D	.20	
Texaco: Louisiana Settlement		
With 3D	.62	.95
Without 3D	.33	.75
Mobil: South Texas		
With 3D		.84
Without 3D		.70

Note: NA is not applicable.

Source: Amoco data are from Aylor 1995a, 1995b, 1996. Exxon data are from Schoenberger 1996. Fairfield Industries data are from Lawrence, Logue, and Grimm 1995 and from personal communication with Marc Lawrence, vice president, Fairfield Industries. Texaco data are from the 1997 Report on the Global Settlement Agreement. Mobil data are from Jeffers, Juranek, and Poffenberger 1993.

marily as a development tool and only recently has come to use it as an exploratory tool.

The value of 3D as an exploratory tool comes not only from drilling fewer dry holes than with 2D, but also from finding resources that 2D would have ignored. An example is cited by Aylor (1996, 76) for an exploratory block in the North Sea. With 2D methods, eight prospects were found with estimated probabilities of success between 22% and 53%, while 3D information placed the probability of success below 10% for six prospects (thus, effectively condemning them) and above 60% for two prospects (thus, making them likely targets).

Exxon's experience, described in Schoenberger 1996, refers to the "geologic" success rate associated with 244 exploration wells in three offshore areas. Geologic success means that hydrocarbons were discovered, though they may not be economic to develop. Geologic success therefore overstates commercial success as used here, and Exxon's numbers in Table 3-1 should be deflated in comparison with those of Amoco. Also misleading is the observation that success rates based on 3D data are about 50% higher than those based on 2D data. Because 3D is more accurate in discerning the presence of commercially valuable deposits, not just geologic presence, the difference between 3D and 2D should be even greater. This is one of the important findings by Aylor, as indicated by the use of 3D to reevaluate 2D prospects.

The Fairfield Industries entries refer to three different samples of observations obtained in 1994 for the purpose of marketing that company's seismic service business. The first entry, described in Lawrence, Logue, and Grimm 1995, is obtained from a survey of twenty-three companies actively exploring in the Gulf of Mexico (out of a total of seventy-one operators in 1994). Thirteen of the companies (57%) responded that they do not use 3D for exploration. The remaining ten companies (43%) said they drilled 104 wildcat wells in the Gulf during 1993 and 1994 using 3D data, and that 53 were successful (51% success rate). No information is presented on success rates with the use of 2D data.

The second Fairfield entry reflects the company's experience with regard to exploration projects in shallow water offshore Louisiana. No information is provided regarding the extent or nature of this experience, though one of the authors indicated that the numbers have not been contradicted when presented before audiences of exploration personnel.[13] The same comments apply to the third entry regarding exploration onshore in south Louisiana.

The Texaco entry refers to experience resulting from implementation of the Global Settlement Agreement, which was the settlement of a dispute between Texaco and the state of Louisiana over natural gas royalty pricing. In addition to a $250 million payment to the state, Texaco agreed to spend $152.25 million over a five-year period starting in 1994 to drill on state leases in south Louisiana. Information regarding these drilling activities is provided to the Louisiana Department of Natural Resources and is monitored by the Center for Energy Studies at Louisiana State University.[14] During 1995, Texaco drilled fourteen wildcat wells, of which eight were drilled with 3D information and six were not; it also drilled thirty-four development wells, of which twenty-two were drilled with 3D information and twelve were not.[15] The resulting success rates are quite high in comparison with the others, which is a reflection of the maturity and low-risk nature of the South Louisiana province.

The Mobil entry, described in Jeffers, Juranek, and Poffenberger 1993, refers to relatively low-risk development drilling in the South Texas Lower Wilcox trend, which provides "a unique opportunity to compare results from like vintages of 3D and 2D seismic data." Sixty-nine wells were drilled during 1991 and 1992, of which thirty-two were based on 2D data and thirty-seven were based on 3D data. Although the historic success rate for development drilling in this area was 70% prior to 1991 based on 2D data, 2D was regarded as inaccurate in imaging the complex subsurface structure. In addition to dry hole savings, the increased accuracy of 3D led to drilling in better locations. For example, the average 3D well brought in 37% more reserves than the average 2D well. After deducting seismic and drilling costs, the net present value of the stream of earnings from 3D wells was double that of the 2D wells. These results demonstrate the value of 3D over 2D even in low-risk development applications.

The Reduction in Exploration and Development Costs

3D seismology affects overall exploration and development costs by increasing surveying costs and reducing the number of wells that need to be drilled. Based on the experiences referred to in the preceding section, surveying costs for 3D are roughly double those of 2D.[16] Exploratory drilling costs decline, on the other hand, because of the increase in the success rate from 20% to 50% per well drilled. Thus, a doubling of seismic costs leads to a 2.5-fold increase in the discovery rate.

The foregoing information is used to calculate an estimate of the impact of 3D seismology on average finding costs. Average finding costs (AFC) may be decomposed into the sum of average drilling costs (ADC) and average nondrilling costs (ANC), all expressed per unit of discoveries:

$$\text{AFC} = \text{ADC} + \text{ANC}$$

AFC is changed as a result of the introduction of 3D by an increase in the discovery rate by a factor of 2.5 and by an increase in the cost of seismic services by a factor of 2. The new AFC, written as AFC*, thus becomes:

$$\text{AFC}^* = (1/2.5)[\text{ADC} + (2t)\text{ANC} + (1 - t)\text{ANC}]$$

where t is the share of seismic surveying costs in ANC. The change in average finding costs is thus:

$$AFC - AFC^* = (1/2.5)[1.5ADC + (1.5 - t)ANC]$$

Dividing through by AFC gives the percentage change in average finding costs due to 3D seismology.

In recent years drilling costs have been about 50% of total exploration costs, while seismic costs are about 40% of nondrilling costs. Inserting these numbers into the above equation puts the percentage change in average finding cost at 0.40 [equal to (0.6)(0.5) + (0.6 – 0.4)(0.5)]. Thus, 3D seismology may be said to reduce average finding costs by 40%.

The same approach may be used to calculate the reduction in the average cost of development drilling made possible by 3D-related increases in success rates. In this case, the improvement in the development success rate from 70% without 3D to 90% with 3D may be said to reduce average development costs by 22%.

Implications for Resource Development

3D seismology increases the amount of the existing resource base that may be exploited at current prices and thus expands the effective size of the resource base. To understand the implications, it is useful to observe that the amount of hydrocarbons physically located in the earth's crust per se is largely irrelevant to the well-being of society. What is important are the expected costs and revenues associated with finding additional hydrocarbons. As long as expected net revenues exceed expected costs, whatever their magnitude may be, exploration and development activity will continue. If expected costs exceed expected revenues, exploration and development will cease, no matter how much oil and gas remain in the earth's surface.

In the absence of technological improvements, expected costs will rise over time as a result of diminishing returns in reserve replacement. Deposits that are the largest and easiest to find will enter the reserve inventory first. Once found, other things being equal, those prospects with lowest development and production costs will tend to be exploited first. Of course, surprises can and do occur because knowledge of the resource base is imperfect, but depletion of the resource base will inexorably push up costs of production unless the means of finding and producing the resource improve.

The advent of 3D seismic methods delays the effects of depletion on costs of production by finding resources that would have been ignored otherwise and by increasing recovery rates from known reservoirs. However, the new technology will not eliminate the depletion effect. Eventually, the cost of adding new reserves must rise as the resource potential susceptible to discovery by 3D methods is depleted. This is the pattern exhibited by earlier exploration technologies, as described in Linseth 1990, where each innovation improved the industry's capability of finding new deposits in particular circumstances. As those opportunities are depleted over time, the rate of new discoveries falls until the technology is replaced and a new set of opportunities is revealed. For example, 3D methods are particularly useful for identifying structural traps caused by the deformation of rock layers. In time, most structural traps will have been identified, and the advantage of using 3D seismology will fade.

HORIZONTAL DRILLING TECHNOLOGY

The second important innovation of the last decade is horizontal drilling. *Horizontal drilling* refers to the ability to guide a drillstring, with a motor at the business end turning the drillbit, that can deviate at all angles from vertical. With this technology, it is possible for the wellbore to intersect the reservoir from the side rather than from above.

The essential differences between vertical and horizontal wellbores are illustrated in Figure 3-6. If formation A is the pay zone, the horizontal well in the middle can achieve greater exposure of the formation to the wellbore than the vertical well on the left. If the formation contains oil or gas bearing fractures as in formation B, the horizontal well on the right is capable of intersecting multiple fractures with the same wellbore while the typical vertical well is limited to a single fracture.

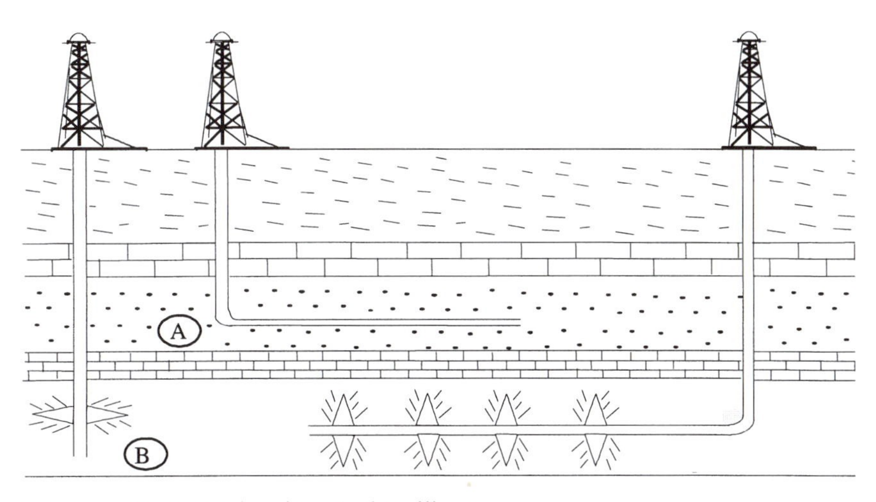

Figure 3-6. Horizontal and Vertical Wellbores.

Note: Area A illustrates the wellbore exposure of horizontal and vertical wells. Area B shows the intersection of fractures with horizontal and vertical wells.

The Benefits and Costs of Horizontal Drilling

Horizontal wells are most advantageous when reservoir conditions call for greater contact between the wellbore and the reservoir formation, as illustrated by formation B in Figure 3-6. One such example is a reservoir that contains a thin pay zone. A vertical well drilled into such a reservoir would encounter a small pay zone and attain a limited capability to draw off oil and gas. In addition to low productivity, a vertical well drilled into a thin pay zone has a high risk of coning, where water penetrates the reservoir and moves toward the wellbore. When the well starts producing water instead of oil, it must be shut down and a new well drilled at a different location in the reservoir. Finally, unless many wells are drilled into a thin reservoir, much of the oil-in-place will not be extracted. Productivity can be improved, fewer wells need be drilled, and more oil can be extracted if the wellbore is guided so that it enters the pay zone horizontally.

A similar advantage occurs with reservoirs in low-permeability rocks. Low permeability means that it is difficult for gas or oil to flow through the interconnections between the pore spaces of the rock and into the well. Horizontal wells can be economic in low-permeability zones when vertical wells are not because of the difference in contact area between the wellbore and the reservoir. This is also true in heavy oil reservoirs, where steam injection is used to drive oil to a wellbore. Horizontal wells serve as high-volume collection points in these applications.

Vertically fractured formations represent another situation where horizontal wells are especially productive. In fact, application of horizontal drilling in the vertically fractured Austin Chalk formation located in South Texas in the 1980s represents its first extensive use in the United States. The formation is an oil- and gas-bearing limestone that extends for many miles. Vertical fractures in the limestone allowed oil and gas to migrate from below the formation up into limestone. Each fracture contains only a modest amount of hydrocarbons and can be accessed only one fracture at a time with conventional vertical wells. A horizontal well, in contrast, can be drilled at an approximate right angle to intersect several vertical fractures, thus multiplying the number of productive zones that can be accessed with the same well.

A final category for which horizontal drilling is well-suited involves re-entry into depleted and abandoned reservoirs. On average, less than 30% of the oil contained in a reservoir has been extracted in the past, because it was uneconomic to produce what remains with conventional vertical wells. With horizontal wells, oil-bearing zones that are too thin, less permeable, or too small to be drilled vertically can become economic again. Moreover, re-entry wells are significantly less expensive than new wells, because they use an existing vertical wellbore, and they benefit from prior knowledge about geologic formations that reduces the risk of a poorly placed wellbore.[17]

The benefits of horizontal wells are not achieved without cost, however. As indicated by King (1993, 4), the cost of a horizontal well can run anywhere from 25% to 400% more than a conventional vertical well, depending on the circumstances. In addition, horizontal wells encounter greater risks than conventional wells and require the use of more sophisticated techniques to offset those risks. One problem with drilling wellbores that curve laterally is the buildup of compression in the drillstring that causes increased contact with the wall of the borehole, leading to higher friction, drag, rotating torque, and lockup. When the drillpipe becomes stuck, the well must be abandoned. The same risk is encountered when the horizontal section of a well traverses certain reactive shales and water is absorbed from the drilling mud, causing the clay minerals in the shale to swell.[18] This action causes the mud and drillpipe to press tightly against the wall of the borehole and may result in a stuck pipe.

The risk that drilling mud can damage the reservoir is also increased. The greater exposure of the reservoir to the wellbore, which is the source of the benefit with horizontal wells, also means there is greater exposure of the reservoir to drilling fluids. When the pressure exerted by drilling mud exceeds the reservoir pressure (that is, overbalanced drilling), the mud will invade the pores of the reservoir and form a hard mud cake that must be cleaned off before production can take place. The risk of damaging the reservoir can be avoided by underbalanced drilling—that is, where the pressure exerted by the drilling mud is slightly less than the pressure in the reservoir—but not without increasing the risk of a blowout. With underbalanced drilling, some oil and gas enters the wellbore and flows to the surface along with the drilling mud. In effect, production is occurring while the well is being drilled. Because the risk of a blowout increases in this circumstance, changes in downhole pressure are monitored closely to avoid serious accidents.

Implications for Productivity

The choice between a conventional vertical well and a horizontal well is one of balancing the cost of developing a reservoir with the amount of reserves added and a higher production flow rate. Horizontal wells are generally more expensive than conventional vertical wells, although in many applications a vertical well may not be a feasible option. For wells drilled into the same formation, horizontal wells cost four to eight times as much as vertical wells as recently as 1989, but by 1993 the cost premium had fallen to a multiple of 1.2 to 1.5. A rough rule-of-thumb used in the industry (*Offshore* 1995; Butler 1988) is that a horizontal well will produce two to five times the rate of output of a conventional well drilled in the same area and, where it does, will typically replace two to five vertical wells. The smaller cost premium, combined with the reduction in the required number of development wells, shifts the cost advantage decisively in favor of horizontal technology in appropriate applications.

It is useful to mention a few individual experiences with horizontal drilling to get a better appreciation of the results. As noted above, early use of horizontal drilling in the United States was concentrated in the Austin Chalk formation of South Texas. The Giddings field, in turn, is the most prolific part of the Austin Chalk formation. This 1,200-square-mile field was discovered in 1961, but its potential was not evident until after 1973, when oil prices rose and extensive field development took place. Field development declined rapidly after 1982, partly because of falling oil prices and partly because of declining flow rates. Interest was rekindled in 1984 with the possibility of drilling horizontal wells that could connect multiple vertical fracture systems with the same wellbore.

Amoco Production Company drilled eight horizontal wells into the Giddings field during 1987–1989 and compared their output rates with the production histories of vertical wells for the same period of time and with equal pressure conditions. As reported by Shelkkholeslami and others (1991), horizontal wells produced between 2.5 and 7 times as much as vertical wells. In another review of 91 horizontal wells drilled in the Giddings field over four Texas counties, Maloy (1992) found that the average well paid back an estimated 60% (discounted at 10%) after-tax return on investment and recovered drilling costs in 1.1 years. As is typical for horizontal wells, the rapid payback is the result of high initial production rates. Equally typical, these small reservoirs are drained quickly and production rates decline rapidly.

The area with the second-most intensive application of horizontal drilling is the Mississippian Bakken formation of North Dakota and Montana. This formation is an oil shale believed to be the original source rock, meaning that the oil is found in the pores of the rock in which it was generated rather than in a conventional trap to which the oil has migrated from a source rock. Johnson (1990) reports that the higher production rates of Pacific Enterprises Oil Company's horizontal wells contributed to a 40% greater return on investment compared to its vertical wells in the same formation.

The third area of concentrated application of horizontal drilling techniques is in the Prudhoe Bay field on the North Slope of Alaska. The field has been in production since 1977 and reached a peak rate of output of two million barrels per day (bpd) in 1988, but by 1995 output had fallen to 1.4 million bpd. Arco's share of North Slope production, in contrast, remained flat at 400,000 bpd during 1985–1995 (Thompson 1996). This level of output was 31% higher than the amount estimated for 1995 ten years earlier. One reason for the difference is the application of horizontal technology. Arco discovered that there were many isolated pockets of oil remaining in the reservoir that could be exploited with horizontal wellbores. Moreover, many of the horizontal wells could be drilled as horizontal sidetracks from existing vertical wells in order to reduce the cost of new reservoir penetrations. If average drilling costs had not declined from earlier levels, Arco figures that there would be few if any wells drilled at Prudhoe Bay today.

Implications for Resource Development

Horizontal wells have important resource development implications because recovery rates and extraction rates are higher than is possible with conventional technology. Higher recovery and extraction rates mean that some development projects become economic with horizontal technology that were not viable with conventional technology. As a result, production occurs from prospects that would have been bypassed or abandoned before and from previously abandoned reservoirs that have been revived. A higher recovery rate helps extend the productive life of known reservoirs by delaying the date in which depletion raises production costs above expected revenues. For these reasons, horizontal drilling technology expands the "effective" size of the resource base compared to the amount that was previously available.

DEEPWATER EXPLORATION AND DEVELOPMENT TECHNOLOGIES

A wide variety of technologies are used to find and develop oil and gas prospects in deepwater (water deeper than 1,000 feet),[19] including drillships, directional drilling methods, production platforms, remote-controlled subsea wells, and subsea pipelines.[20] One of the most prominent features of deepwater exploration and development is that it is very expensive. In recent years, outlays per project in the Gulf of Mexico have been running up to a billion dollars and more. The investment has been profitable, even though petroleum prices have been stagnant since 1986, because of high success rates and because reservoirs in the Gulf of Mexico have proven to be highly productive.[21]

Exploratory drilling in deepwater is accomplished using either a semisubmersible drilling platform (for water depths up to 2,000 feet) or a drillship (for deeper water).[22] A *semisubmersible* is a platform connected to a set of flotation pontoons, where most of the flotation is positioned below sea level in order to minimize motion even in rough seas. A drill-ship has a drilling rig mounted amidships and a control system for maintaining position over a drillsite. If a commercial field is discovered, it may be developed using one of a variety of production platforms. In shallower water (less than 1,500 feet in depth), a steel jacket platform may be used that has steel legs that are piled into the seabed.

In water of moderate depth (between 1,500 and 5,000 feet), a *tension-leg platform* (TLP) or other floating production system is used. The TLP floats above the offshore field and is anchored to the sea floor by hollow steel tubes called *tendons*. The tendons pull the platform down in the water to prevent it from rising or falling with wave action (hence, tension legs). The advantage of platforms is that production wellheads and processing equipment are located on the platform. The on-board processing facilities separate oil from gas and water and clean the produced water for disposal in the ocean. The oil and gas are then transported to shore via pipelines for sale.[23]

In very deep water (beyond 5,000 feet), subsea completions may be the preferred option. Subsea wellheads located on the ocean floor are connected to an underwater pipeline system to carry the oil and gas to shallower water, where a platform is located that can handle processing and shipment by pipeline to the shore. Subsea satellite systems are cheaper than platforms when production centers are dispersed but, unlike platforms, require mobilization of a costly floating vessel to support well re-entry for maintenance. If wellheads and pipes become plugged with sand, paraffin, or hydrates, they must be cleaned out to maintain production. Cleaning usually requires re-entry from the surface, which is especially costly when subsea wellheads are many miles from a platform.[24]

Development wells drilled from production platforms commonly use directional drilling techniques. Directionally drilled wells may fan out from the drilling platform in all directions to reach one or more reservoirs located several thousand feet from the platform. A strategically positioned platform can be used to develop separate reservoirs located over a wide area. For example, leaseblocks in the Gulf of Mexico are on average about five square kilometers, making it possible with current technology to drill into as many as twelve surrounding leaseblocks from the same central platform.

Implications for Productivity

Without the recent advances in technology that lower costs, reduce risks, and extend the capability of the industry, deepwater reserves would remain out of reach. The three technologies discussed above all contribute to the practicality of deepwater exploration and production. High-quality seismic information is needed to determine the probability of finding oil and gas reserves with sufficient accuracy to justify bidding for a lease and investing in exploratory drilling. Drilling and production could not proceed without the development of platforms, drillships, and subsea systems capable of operating in deepwater. And expensive platforms would not be practical without the ability to drill directional wells to targets located at extended distances from the platform.

Figure 3-7 illustrates the progress in offshore production capability by listing the depth records achieved in the Gulf of Mexico. The ability to explore in ever deeper water increased steadily since offshore production began in 1947, though truly deepwater production did not begin until the introduction of the tension-leg platform in 1989.[25] In 1988, Shell com-

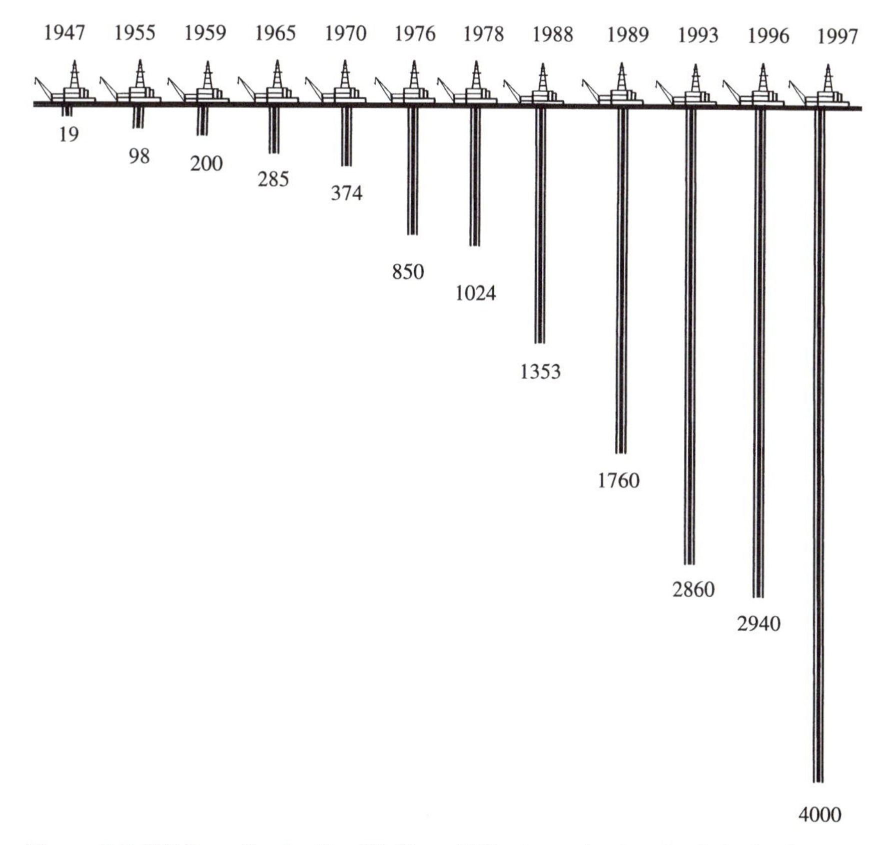

Figure 3-7. Offshore Production Platform Milestones (water depth in feet).

Source: Shell Briefing Service 1993; Wheatley 1997.

pleted the Bullwinkle platform in the Gulf of Mexico, the deepest fixed-leg platform in the world at 1,353 feet and considered to be the depth limit for fixed-leg platforms. Shell also set the 1997 depth record with the Ursa TLP placed in 4,000 feet of water. In just nine years, the depth capability increased by 2,647 feet, or by nearly twice the entire amount achieved in the previous forty years.

For comparison with the depth of platforms, the 1997 world depth record for petroleum production is 5,000 feet, set by Shell's Mensa field located about 140 miles southeast of New Orleans. Unlike platform systems, where the wellheads are located above sea level, the Mensa field uses three wellheads located on the sea floor. These wells are tied back to an existing shallow water platform using a sixty-eight-mile pipeline. The tieback to a distant platform is possible in this case because the gas is "dry." More common wet gas would crystallize at these depths and limit the tieback distance.

Craig and Hyde (1997) have estimated the rates of return on investments in three deepwater projects: the Auger TLP in 2,871 feet of water, the Mars TLP in 2,940 feet, and the Mensa subsea wells at 5,000 feet (all developed by Shell). Under fairly conservative assumptions about costs, revenues, and discovery sizes (such as a flat real price of oil of $18 per barrel and real price of gas of $1.80 per thousand cubic feet), the after-tax rate of return comes to 14.8% for Auger, 29.8% for Mars, and 27.5% for Mensa. Actual profit rates may be expected to be much larger because early estimates of reserve volumes are generally very conservative. In view of the high returns, it is little wonder that the industry is now offering record bids for deepwater leases.

The movement into deeper water has significantly raised the productivity of the U.S. petroleum industry, as measured by exploratory success rates and output per well. The average success rate for exploratory drilling in the Gulf of Mexico is about twice that for the United States in general. In addition, output per well in the Gulf of Mexico is significantly higher in deeper water than in shallow water, contributing to a significant improvement in the payback period for investments in deepwater. Wells located in water less than 500 feet generally produce less than 200 barrels of oil equivalent per day, while wells in water deeper than 500 feet are producing in excess of 1,000 barrels per day, a fivefold difference.[26]

ORIGINS AND IMPLICATIONS OF THE TECHNOLOGIES

Now that the productivity implications of the three groups of technologies have been established, we wish to consider where these innovations came from and where will they lead. What technical barriers were overcome in course of their development to make them successful? How quickly were the innovations adopted by the industry? Were the technical advances induced by the profit incentive, or did they developed independently? With regard to the implications of these innovations, what is their contribution to the competitiveness of the U.S. petroleum industry and their effect on the world price of oil?

Origins of the Technologies

3D Seismology. The theoretical principles of 3D seismology have been known for many years and the technology was used as early as the 1970s, but practical success and widespread adoption had to wait until the development of parallel computing in 1985. Without increased computing power, and the corresponding reduction in the time and expense of using 3D methods, this new technology would have been relegated to a specialty role in exploration and development. Once the ability to process vast amounts of data was achieved, better ways of recording and collecting the data (such as multistreamer, multisource, multivessel marine technology), transmitting the data (such as satellites), and using the data (such as workstation technology, analytical algorithms, and geophysical models) soon followed.

The development of high-performance computing was closely associated with the needs of the petroleum industry. According to the National Research Council (1996, 1), "Seismologists were among the first scientists to exploit the capabilities of advanced computing technology." There was a ready demand for high-speed computers in the petroleum industry.

The difference between conventional digital computers and parallel computers is straightforward. Conventional computers consist of a memory for holding data and a processor for transforming data, where transfers of data back and forth between memory and processor take time and limit processing speed. Parallel computers, in contrast, integrate the memory and processor functions by distributing memory physically throughout the computer to places where it will be needed. Hillis (1992) compares the technology of parallel computing to traditional mass production technologies in industry, where economies of scale are exploited by doing more than one thing at the same time. The performance of similar operations on many elements of data at the same time is responsible for the massive increase in processing speed.

The number of concurrent processing operations can be increased by adding more processors. Where a conventional computer takes twice as much time to process twice as much data, a parallel computer can process twice as much data in the same amount of time simply by doubling the number of processors. The reduction in the time required to process seismic data with parallel computers is dramatic. As indicated by Nestvold and others (1996), the computer time required to process one square kilometer of seismic data declined from more than 800 minutes in 1985 to less than ten minutes by 1995. The increased speed means that seismic analysts can process data almost as fast as surveyors acquire it, so that promising results can be identified and resurveyed almost immediately.

Horizontal Drilling Technology. Experiments with horizontal drilling techniques date back many decades, but economic success depended on the development of three complementary technologies: 3D seismic information, so drillers would know where they wanted to place the borehole; steerable motor assemblies, so that drillers could guide the drillstring to the target; and a downhole sensor package, so that drillers would know where the drillstring was located relative to the target. Dedicated computers are used to compare expectations about subsurface structures obtained from seismic information with actual experience obtained during drilling to aid in decisions about the drilling strategy.

Much was said earlier about the gains achieved with 3D seismology, but little about the antecedents to steerable downhole motors and downhole sensory packages. Before steerable motor assemblies made horizontal drilling more practical, bent subs were used on downhole motors to direct the drillstring. A *bent sub* is a short section of pipe with an angle machined into it that, when lowered into the borehole to the desired depth, forces the drillstring to move off at an angle. These devices were a major advance, in turn, over earlier rotary drilling assemblies, but they are inaccurate, time-consuming, and expensive to use. Frequent "trips" are required to change bent subs in order to control the placement of the well, partly because the angle they produce is either too sharp or too broad and has to be continually adjusted. Steerable motor assemblies improved directional control and reduced trip time because one motor assembly could be used for the entire directional portion of the well.

Finally, before measurement-while-drilling (MWD) packages, single-shot surveys were used to judge location of the drill bit. With the older technique, survey tools are run down the hole near the drill motor, where magnetic interference from the drillstring reduces accuracy. Electronic MWD instruments represented such a radical departure from the old methods that they were not trusted at first. Experience soon proved that MWD tools greatly improve directional control, enabling drillers to intersect smaller targets and to achieve more accurate placement of the wellbore in the pay zone.[27]

Deepwater Systems. The industry's capability to produce oil and gas from deposits located beneath water developed gradually from 1947 to

1988 and increased dramatically thereafter. Although many separate technologies were involved, three pivotal innovations mark the turning point in the industry's capability to operate in deepwater: 3D seismology, directional drilling, and production platform design. 3D seismology was important to deepwater exploration because it provided more accurate estimates of reserve potential that were required to justify the enormous capital investments needed for deepwater exploration, development, and production facilities. Directional drilling enables operators to drill multiple production wells from the same production platform, thus increasing the productivity of individual platforms and reducing the number of platforms required for field development. The third major technical advance in deepwater development, the tension-leg platform, greatly extended the reach of the industry into frontier areas.

Diffusion of the Technologies

3D Seismology. The rate of adoption of 3D seismic methods within the industry has been inversely correlated with the cost of using the technology and directly related to the benefits gained from improved information. The increased accuracy of 3D images was appreciated from the beginning, but the high initial cost of using the technology was responsible for relegating it to a development role at first. As costs declined over time, the technology began to be applied in an exploratory role. With the growing record of experience in this dimension, the full benefits of 3D methods began to appear.

3D technology is as widely available in the rest of the world as it is in the United States, but perhaps is not as widely adopted. Access to the technology is assured even if host governments and national petroleum companies abroad do not have the expertise to put it into practice. The technology may be transferred through arrangements with international petroleum companies, who are awarded local concessions or leases, or through contracts with seismic service companies. The latter operate worldwide and are no less inclined to offer their services abroad than in the United States.

Horizontal Drilling. The speed with which horizontal drilling techniques are adopted within the United States seems to be company-specific and reservoir-specific. The advantages of horizontal drilling are greater for oil deposits than for gas deposits simply because reservoir conditions are more likely to impede the flow of oil to the borehole. Regionally, horizontal drilling first took off in the Austin Chalk formation of Texas, then in the Bakken Shale formation of North Dakota, and most recently in the Prudhoe Bay field in Alaska. In each case, reservoir characteristics make horizontal drilling particularly advantageous. Naturally, companies located in these areas are among the leaders in developing horizontal drilling techniques, the most skilled in using the technology, and the most aware of the advantages to be gained.

Like 3D seismology, access to the latest horizontal drilling technology and the skills needed to use it can be purchased from specialized service companies. What cannot be purchased as easily is knowledge about the efficacy of the technology in new situations and a corporate willingness to try new approaches. The same considerations slow the transfer of this technology to other countries.

Deepwater Development. The deepwater technologies have been developed by only a few firms ready to make the necessary financial commitment. The world leader in this technology is Shell Oil Company.[28] Other companies have been investing in deepwater exploration, to be sure, but the successes of the leaders have been responsible for the recent stampede in this direction.

Motivation for the Technologies

Depending on one's perspective, it is possible to conclude that the profit motive was of more or less importance to the development of the technologies discussed here. As noted earlier, lower oil and gas prices and higher production costs provided powerful incentives to seek better ways of finding and developing oil reserves. While true, it is perhaps an oversimplification to say that these technologies bloomed in the last ten years solely as a result of concerted efforts by the industry to develop them. The process has been much longer, much more complex, and much more serendipitous than the direct profit motive might imply.

All three technologies were on the conceptual drawing board many decades before the commercial successes achieved in the last decade. In addition, all three technologies were subjected to decades of testing various adaptations, each of which represented a small marginal improvement over the last iteration. Finally, all three technologies experienced a quantum leap forward in the last decade. Thus, while these technologies were all commercially proven in the 1990s, their success was the result of small incremental improvements over many years and a recent momentous change. The recent momentous change, moreover, is linked to the largely exogenous development of high-speed computers: they provided the essential breakthrough for 3D, and 3D provided the essential breakthrough for both horizontal drilling and deepwater development. In short, on one level at least, the innovation process does not seem to be directly influenced by incentives to reduce costs of production as motivated by the decline in oil and gas prices in the early 1980s.

It would also be a mistake to underestimate the extent to which endogenous influences were responsible for the development of these technologies. None of them suddenly appeared in the last ten years in readily usable commercial form. In fact, all three technologies are better described as groups of complementary technologies that support a central concept. Thus, for example, high-speed computers were the essential breakthrough necessary to implement the concept of 3D, but this event by itself was not sufficient for success. Computers had to be told how to process the data, and methods had to be developed for understanding what the processed data meant.

The process of creating complementary groups of technologies involves, in addition, repeated field testing to identify the weak links that require improvement. During this phase of activity, setbacks can be formidable barriers to continued development and require the dedicated commitment of firms to overcome them. For example, early applications of horizontal drilling on the North Slope of Alaska in the mid-1980s encountered serious problems with stuck pipe and blowouts, and it was decided to halt further drilling until the problems could be solved. When drilling resumed in 1990 with new equipment and drilling strategies, the problems were avoided and the payoff from horizontal techniques began to emerge.

The development of a new technology is expensive and, until proven, the benefits to be gained are far from certain. Nevertheless, the industry forged ahead in these three important areas and in just a few years achieved remarkable successes. The commitment to develop, test, and improve the technologies can be attributed to the economic incentive to lower the costs of finding and developing reserves. Progress also depends on the existence of firms willing to take on a leadership role, such as, for example, Amoco and Exxon with 3D, Arco and British Petroleum with horizontal drilling, and Shell and Petrobras with deepwater systems.

Technical success does not alone ensure that innovations will be widely adopted in the industry. Additional time is required before the innovators can establish a record of experience that demonstrates the economic superiority of new relative to old technology. The remarkably short period of time required for these technologies to diffuse throughout the industry attests to their economic superiority. Virtually every issue of *Oil and Gas Journal* and *Offshore* in the last few years carried at least one article announcing another successful application of one of the three technologies discussed here. It is little wonder that the technologies caught on quickly once the success stories started to pour in, but one must keep in mind that innovators had to be there first to create the stories.

Implications of the New Technologies

Competitiveness of the U.S. Petroleum Industry. The introduction of 3D seismic methods, horizontal drilling, and deepwater production technologies have all improved the competitive position of the U.S. petroleum industry in the world market, particularly with respect to low-cost producers. The comparison with low-cost producers is pertinent because production decisions in these countries determine how much of the world market will be shared with high-cost producers.

3D technology, in particular, has significantly reduced the cost of finding new reserves. Although the same technology is equally available in the rest of the world, the impact on profit margins is more important in high-cost areas such as the United States than in low-cost areas such as the Persian Gulf. For example, a 40% reduction in finding costs is important when finding costs are several dollars per barrel and not important when they amount to no more than a few cents per barrel. Additionally, the ability of 3D to locate new deposits is less important in prolific areas such as the Persian Gulf than in less well-endowed environments such as the United States. Thus, 3D technology helps the competitive position of the petroleum industry in the United States, as well as in other high-cost countries.

Horizontal drilling and deepwater technologies also help bolster U.S. profit margins. Both technologies extend the effective size of the domestic resource base that is economic to exploit. Although both technologies are available anywhere in the world, they do not have the same impact on profits in many areas abroad as they do in the United States. In particular, horizontal drilling and deepwater technologies are important in the North Sea, Brazil, West Africa, Southeast Asia, and the Caspian Sea, but not in the Persian Gulf. Horizontal drilling is less important in the Persian Gulf because the resource base has not been depleted to the point where it is worthwhile to re-enter or rework older fields. Higher earnings can be achieved by putting investments into new reservoirs or into developing existing reservoirs more fully. Also, deepwater technology is not applicable in the Persian Gulf because of shallow waters.

Implications for the World Price of Oil. Although 3D seismology, horizontal drilling, and deepwater technologies lower production costs and raise domestic output in the United States, these technologies are not likely to have a major, immediate impact on the world price of oil. This conclusion is perhaps self-evident in the case of horizontal drilling and deepwater production because they do not affect a large share of world production. 3D seismic technology is another story because it has a dramatic effect on finding costs worldwide. Nevertheless, the expected impact on the price of oil will be small and possibly hard to detect in recorded price data. The reason is that the induced reduction in finding costs is a small fraction of the world price.

To illustrate, suppose that 3D results in a 40% reduction in average finding costs worldwide, not just in the United States. Since finding costs average between $4 and $5 per barrel, according to estimates by Arthur Anderson (1994), it follows that 3D can reduce total costs by as much as $2 per barrel. The potential reduction in price would therefore be, at most, $2 per barrel.[29]

For comparison, during 1986–1995 the average monthly world price fluctuated about a mean of $18.30 with a standard deviation of $5.77.[30] Thus, a $2 per barrel reduction in the price amounts to slightly over 10% of the average world price and about a third of the standard deviation. The magnitude of the effect of 3D seismology on the price is therefore modest relative to the many other market factors that cause fluctuations in the world price.

While 3D seismology may not cause a noticeable drop in the price, the contribution to lower costs will keep the price from rising as much as it otherwise would in the future. Also important is the potential effect of 3D on future price variation. The reduction in finding costs, as noted, is more important to high-cost producers than to low-cost producers. The low-cost producers are dominated by the members of OPEC (Organization of Petroleum Exporting Countries), and especially by those located on the Persian Gulf. The high-cost producing countries, in contrast, are a collection of independent producers with relatively small levels of output. This group of producing countries is sometimes characterized as the "competitive fringe" in the oil market. Their presence limits the effect that changes in production by the dominant countries will have on the market price. For example, a reduction in output among the dominant countries will raise the market price, but the higher price will lead to an increase in the amount of output from the competitive fringe. Conversely, an increase in output elsewhere will force a reduction in output from the competitive fringe. Thus, output from the competitive fringe tends to move in the opposite direction of output from the dominant countries. In this way the high-cost producers play a price-stabilizing role in the market, and new technology expands the role of the high-cost producers.

ACKNOWLEDGMENTS

Special thanks go to Bill Aylor of Amoco, Paul Erb of Conoco, Allan Pulsipher of Louisiana State University, Bill Fisher of the University of Texas at Austin, and

David Morehouse of the Energy Information Administration for reading and commenting on an earlier draft of this paper. Other people who were especially helpful in providing information are Ed Reynolds and Ed Parma of Conoco, Inc.; Bob Rothberg of Amoco Corporation; Mike Schoenberger of Exxon Production Research Co.; Don Armijo of Shell Offshore Inc.; Paul Versowsky of Chevron Co.; Bill Hill of BP Exploration (Alaska), Inc.; Marc Lawrence of Fairfield Industries; Rick Hahn of OPE, Inc.; Phil Schearer of Western Geophysical; James Ragland of Aramco Services Co., and Rob Haar and Bob King of the Energy Information Administration. Thanks are due to Shell Oil Company for inviting me to visit their Bullwinkle Platform in the Gulf of Mexico and to Arco Alaska Inc. and BP Exploration Alaska for inviting me to visit their facilities at Prudhoe Bay. A special debt of gratitude goes to Brian Kropp, who prepared the figures and calculations, proofread drafts, and reduced the number of mistakes in the manuscript.

ENDNOTES

1. Added to the depletion effect is the rising cost of actions taken to safeguard the environment. Over the last twenty years, firms have been required to meet increasingly stringent standards for controlling air and water pollution, and as a result they have had to adopt more expensive practices for mitigating the effects of their activities on the environment. In effect, the costs of safeguarding the environment that were once ignored by firms have now been internalized into their profit statements.

2. Such measures follow a pattern that is inversely related to the amount of drilling activity. Productivity declined during 1973–1981, when the price of oil and the level of drilling were high, and increased after 1986, when the price of oil and level of drilling declined. This scale effect on productivity is the result of moving into lower-quality prospects among the inventory of those available to be exploited.

3. Revisions may also result from exploratory (delineation) drilling or from factors unrelated to drilling (for instance, a price increase).

4. In addition to reflection surveys, gravity surveys and magnetic surveys are also used in the search for hydrocarbon deposits. With a gravity survey, a measuring instrument is passed over the earth's surface to measure minute variations in gravitational force. High gravity measurements indicate the presence of high-density rock that has been raised nearer to the surface, such as a buried anticline, while a low gravity measurement indicates the presence of a low-density rock, such as a salt dome. Magnetic surveys, in contrast, look for distortions (from normal) in the earth's magnetic field to identify magnetic anomalies that might be caused by arches and other structural traps in overlying sediments. Magnetic surveys are not considered very reliable as a primary exploration tool but are used as a supplement to seismic and gravity surveys.

5. A good description of 3D seismic methods for nonspecialists is Gadallah 1994. A brief overview is given in Haar 1992.

6. The acoustic source on land is produced by explosives or, in more sensitive areas, a truck-mounted vibrator. Marine sound sources are air guns that displace water volumes.

7. Harder rock layers have higher acoustic impedance and reflect more energy back to the surface.

8. Another important variable for resolution is the distance between the receivers.

9. The recorded data are referred to as *noise* if the data contain no useful information about subsurface structures. *Signals* contain useful information to make inferences about subsurface structures.

10. A survey of subsalt prospects as of 1997 is given in Wheatley 1997.

11. *Coning* refers to the penetration of water or gas to the wellbore of an oil well, causing the well to produce water or gas rather than oil. Gas production may be unwanted because it is the source of natural pressure for extracting oil or because the well is located too far from a gas pipeline system to make it possible to market the gas. When a well cones water, it must be shut down and redrilled.

12. The trade press is replete with other examples of the cost-effectiveness of 3D surveys over 2D, many of which describe how 3D saved money by avoiding unnecessary drilling or increased earnings by showing the location of commercial deposits that were ignored by 2D.

13. Conversation with Marc Lawrence, Vice President of Fairfield Industries, in Houston, Texas, August 1996.

14. The author thanks Allan Pulsipher, Director of the Center for Energy Studies, for providing information from the 1995 report on the Global Settlement Agreement.

15. Of the eight wildcat wells drilled using 3D, five were successful, and three contained oil while two contained gas. For comparison with the six wildcats drilled without 3D, two were successful and both were gas. The successful development wells were predominately oil wells: fifteen out of twenty-one with 3D, and seven out of nine without 3D.

16. The cost premium of 3D relative to 2D has declined in recent years. In a survey of the industry by Bruch and Bohi (1993), it was concluded that 3D was three times as costly as 2D.

17. In at least one reservoir development project in the Gulf of Mexico, the cost of re-entry averaged 37% of the original development wells drilled from the same platform. See *Offshore* 1995.

18. Drilling mud is a mixture of clay and either water or oil that is pumped down the hollow drillstring and through the drillbit to the bottom of the borehole. The mud performs several vital functions. It cools and lubricates the drill bit; powers downhole motors; picks up rock chips from the bottom of the well and transports them to the surface; prevents water from migrating into the borehole; and controls downhole pressures to prevent blowouts.

19. This is the definition of deepwater used by the Minerals Management Service.

20. Within the United States, these technologies are used almost exclusively in the Gulf of Mexico, since they are either banned in other offshore areas (as in California and Florida), have no bright prospects (as in New England), or are shallower (as in Alaska). The Gulf of Mexico served as the world's birthplace for offshore petroleum exploration beginning in 1947, but now accounts for only about 5% of total world offshore production.

21. Evidence to this effect is found in the offshore lease sale conducted in April 1996, where records for the number of bids offered and the number of tracts receiving bids were set. Among the 924 tracts receiving bids, 442 were located in more than 1,350 feet of water, compared to 178 in the previous sale in April 1995 (*Oil and Gas Journal* 1996).

22. Drillships are currently able to drill in water up to 7,500 feet in depth, though several are under construction with a 10,000 foot capability.

23. In other countries, tankers are used for storage and for shuttling oil to shore, though not in the United States because of environmental regulations. As fields become deeper and farther from shore, the cost of pipelines increases, and the restrictions on the use of tankers will become more controversial.

24. When subsea wells are the only option, it may be possible to employ a so-called *mini floater*, which is essentially a small, single-column TLP that can handle only a small number of wells and processing facilities. This concept has not been employed yet, however.

25. Conoco placed the first tension-leg platform in service in the North Sea in 1984, but in shallower water (500 feet) to prove the concept.

26. Part of the difference in productivity is due to the higher age of wells in shallower water. They have been producing longer and are on average more depleted than wells in deeper water.

27. Directional control is maintained by tying the location of the drillstring to geologic markers in rock layers and to oil–water and gas–oil contacts.

28. As of 1996, Shell ranks number one among operators in the Gulf of Mexico in terms of acreage held, producing acreage, wells drilled, and production, both in deepwater and in all depths. See *Offshore* 1997.

29. The maximum price effect assumes that the price elasticity of supply of oil is infinitely elastic and that the price elasticity of demand is zero. A supply elasticity less than infinity or a demand elasticity less than zero will creating offsetting effects following the reduction in finding costs that will reduce the price effect to less than $2 per barrel.

30. The price series is the average monthly F.O.B. cost of imported oil, as reported in Energy Information Administration 1996.

REFERENCES

Adelman, M.A. 1997. What Do Recent N. Sea Unit Cost Changes Mean? *Oil and Gas Journal* 95(5, February 3): 45–46.

American Petroleum Institute. 1996. *Basic Petroleum Data Book*. Washington, D.C.: American Petroleum Institute.

Arthur Anderson. 1994. *Oil and Gas Reserve Disclosures*. New York: Arthur Anderson, Inc..

Aylor, William K. 1995a. Business Performance and the Value of Exploration 3D Seismic. *Geophysics: The Leading Edge* 14(7, July): 797–801.

———. 1995b. Business Impact and the Value of Exploration 3D Seismic. September 15. Unpublished paper. Houston, Texas: Amoco Corporation.

———. 1996. The Business Impact and Value of 3D Seismic. Paper presented at the Offshore Technology Conference, Houston, Texas. May.

Bruch, Vicki L., and Douglas R. Bohi. 1993. The Impact of 3D Seismic Technology on the Petroleum Market. Sandia Report: SAND92-1613. Albuquerque, New Mexico: Sandia National Laboratories.

Butler, R.M. 1988. The Potential for Horizontal Wells for Petroleum Production. Paper presented at the 39th Annual Meeting of the Petroleum Society of CIM, Calgary, Canada. June.

Craig, Michael J.K., and Steven T. Hyde. 1997. Deepwater Gulf of Mexico More Profitable than Previously Thought. *Oil and Gas Journal* 95(10, March 10): 45–48.

Energy Information Administration. Various years. *U.S. Crude Oil, Natural Gas and Natural Gas Liquids Reserves Annual Report*. Washington, D.C.: Energy Information Administration.

———. 1996. *Annual Energy Review 1995*. Washington, D.C.: Energy Information Administration.

Gadallah, Mamdouh R. 1994. *Reservoir Seismology: Geophysics in Nontechnical Language*. Tulsa, Oklahoma: Pennwell Books.

Haar, Robert. 1992. Three Dimensional Seismology—A New Perspective. *Natural Gas Monthly*. 92(12, December): 1–11.

Hillis, W. Daniel. 1992. What Is Massively Parallel Computing and Why Is It Important? *Daedalus* 121(1): 1–15.

Jeffers, Patricia B., Thomas A. Juranek, and Michael R. Poffenberger. 1993. 3D versus 2D Drilling Results: Is There Still a Question? In *Proceedings: 63rd Annual Meeting of the Society of Petroleum Geologists*. Washington, D.C.: Society of Petroleum Geologists.

Johnson, Sandra. 1990. Drillbits Test Wash's Horizontal Potential. *Western World Oil* 47(11): 27–31.

King, Robert F. 1993. Drilling Sideways: A Review of Horizontal Well Technology and Its Domestic Application. DOE/EIA/TR–0565. Washington, D.C.: Energy Information Administration.

Lawrence, Marc A., Hugh T. Logue, and Don A. Grimm. 1995. Reducing Dry Hole Risk with 3D Seismic Data. Report. Houston, Texas: Fairfield Industries.

Linseth, Roy O. 1990. The Next Wave of Exploration. *Geophysics: The Leading Edge of Exploration* 9(12, December): 9–15.

Maloy, William T. 1992. Horizontal Wells Up Odds for Profit in Giddings Austin Chalk. *Oil and Gas Journal* 90(7, February 17): 67–70.

National Research Council. 1996. *High-Performance Computing in Seismology*. Washington, D.C.: National Academy Press.

Nestvold, E.O., C.B. Su, J.L. Black, and I.G. Jack. 1996. Parallel Computing Helps 3D Depth Imaging Processing. *Oil and Gas Journal* 94(44, October 28): 35–44

Offshore. 1995. Horizontal Drilling Makes U.S. Gulf Re-Entry Attractive. 55(2, February): 30–32.

———. 1997. 64 U.S. Gulf Deepwater Prospects in Study: Survey Shows Total of 82 Fields and Prospects beyond 1,000 Ft Depth Contour, of Which 18 Are Already in Production. 57(1, January): 48.

Oil and Gas Journal. 1996. What Fueled the Bidding in Record OCS Sale 157? 94(19, May 6): 40.

Schoenberger, M. 1996. The Growing Importance of 3D Seismic Technology. Paper presented at the Offshore Technology Conference, Houston, Texas. May.

Shelkkholeslami, B.A., B.W. Schlottman, F.A. Seidel, and D.M. Button. 1991. Drilling and Production Aspects of Horizontal Wells in the Austin Chalk. *Journal of Petroleum Technology* 43(7, July): 773–79.

Shell Briefing Service. 1993. *The Offshore Challenge*. London: Shell International Petroleum Company, Ltd.

Thompson, Ken (President of Arco Alaska, Inc.). 1996. Speech before the Anchorage Chamber of Commerce. April 29.

Wheatley, Richard. 1997. Deepwater, Subsalt, and Prospects Open New Era for Gulf of Mexico. *Oil and Gas Journal* 95(3, January 20): 48.

4

Innovation, Productivity Growth, and the Survival of the U.S. Copper Industry

John E. Tilton and Hans H. Landsberg

Productivity isn't everything, but in the long run it is almost everything. A country's ability to improve its standard of living over time depends almost entirely on its ability to raise its output per worker.
—Paul Krugman (1994, 13)

Mining in general and copper mining in particular go back a long way, back to the Bronze Age, possibly even the Stone Age. The Romans mined copper in Spain and tin in England some two thousand years ago. The writings of Agricola in the sixteenth century show that mining was important in the Middle Ages and that Medieval Europe made significant contributions to the art of mining and metal processing. Today, the world mines more mineral commodities in greater tonnages than ever before.

This long and fascinating history, so closely tied to the economic development of the human race, is one of the charms of the mining industry. It also, unfortunately, gives rise to some widespread misconceptions about mining. Old industries are generally considered stodgy, with mature and stagnant technologies. Costs do not decline as in the newer, high-technology industries and may even increase with real wages. Costs

JOHN E. TILTON is William J. Coulter professor of mineral economics at Colorado School of Mines and university fellow at Resources for the Future. HANS H. LANDSBERG is senior fellow emeritus at Resources for the Future.

are expected to rise particularly in the case of mining, as the depletion of the best mineral deposits causes labor productivity to fall (Young 1991).

In fact, however, the mining of copper, and other metals as well, is a highly competitive global industry, where the successful firms aggressively pursue new technologies and other cost-reducing innovations. Costs over the longer term have fallen substantially—indeed, more than the production costs of nonextractable goods—despite the increasing need to exploit lower grade, more remote, and more-difficult-to-process deposits (Barnett and Morse 1963; Tietenberg 1996, chapter 13).[1]

In the pages that follow, we examine changes in productivity over the past several decades in the U.S. copper mining industry along with their causes and consequences. Such an inquiry seems worthwhile given the widespread misconception that mining is a mature industry with relatively static technology. Copper mining in the United States is of particular interest as it has enjoyed in recent years a remarkable revival, largely the result of a dramatic jump in productivity, following what many observers predicted was a terminal decline.

Though the scope of the analysis is limited to the U.S. copper industry, the implications that flow from the findings extend to other mineral commodities and to other countries. The focus is on the first two stages of copper production, namely mining and milling, though it is not always feasible to exclude downstream processing at the smelting and refining stages. For the most part, the time period covered runs from 1970 to the mid-1990s, though in a few instances data back as far as 1950 are noted.

While other studies in this volume concentrate on multifactor productivity, the data needed to measure multifactor productivity are not available for the copper industry. As Parry demonstrates in Chapter 6 of this volume, one can estimate multifactor productivity on the basis of data for all the metal mining industries in the United States, though, as he notes, the validity of the results depends on a number of strong assumptions. This chapter, as a result, focuses mostly on labor productivity, though it does consider Parry's estimates of multifactor productivity in assessing the role that capital and intermediate goods played in the recovery of the U.S. copper industry. As we will see, labor productivity and multifactor productivity have moved in the same direction over the period examined, both falling in the 1970s and rising in the 1980s and 1990s. Multifactor productivity, however, fell more and then rose much less than labor productivity.

The first section following this introduction provides a brief description of the U.S. copper mining industry, focusing mostly on its decline and subsequent recovery. The second section explores possible explanations for the recovery and highlights the importance of the rapid surge in productivity that occurred during the 1980s. The third section attributes this rise in productivity largely to new technologies and other labor-saving innovations. The fourth section looks in some depth at the development and adoption of one important innovation—the solvent extraction–electrowinning (SX-EW) process. The fifth and final section examines the implications of the findings for mineral-producing countries striving to increase their comparative advantage in an increasingly competitive global economy and for all countries trying to raise their living standards by accelerating productivity growth.

DECLINE AND REVIVAL

The United States was the world's largest producer of copper in 1970, as it had been throughout the twentieth century. In that year, as Table 4-1 indicates, its mines produced some 1.56 million tons of copper, accounting for 30% of total western world production. At the mining and milling stages alone, it employed 37,000 people. The industry as a whole was quite profitable. Adjusted break-even costs (as defined in Table 4-1, note d) averaged $1.04 in real (1990) dollars, considerably below the prevailing copper price of $1.83, again measured in real (1990) dollars. Net imports of copper in concentrates, blister, and refined metal supplied only 8% of the total U.S. market.

The fifteen years that followed, however, threatened the industry's very survival. Mine output fell by nearly a third. The country's share of western world output shrank from 30% to 17%. Net imports rose. Adjusted break-even costs, though they did decline, did not fall as much as the price of copper. As a result, few producers were recovering their full production costs, including the fixed costs of capital. Many were not even covering their variable or cash costs. In these depressed conditions, a number of mines curtailed production or shut down completely. Employment in copper mining and milling fell to 13,000, a 70% decline from its 1970 level.

Producers twice petitioned the government for protection under the Trade Act of 1974, first in 1978 and then again in 1984. Both times their request was denied. The government maintained that protection would cost more jobs in the copper fabrication industry than it would save in copper mining and processing. Skeptics noted that unlike the steel industry, which did receive protection, the copper industry was concentrated in states with relatively little political influence, such as Arizona, Utah, New Mexico, and Montana.

Nor was it just the industry seeking government assistance that saw a bleak future for domestic copper mining. The media also were pessimistic. In the mid-1980s, *Business Week* in a cover story declared the death of mining in the United States.

Table 4-1. Mine Output, Share of Western World Output, Net Imports, Adjusted Breakeven Costs, and Prices for the U.S. Copper Industry, 1970, 1975, 1980, 1985, 1990, and 1995.

Year	*1970*	*1975*	*1980*	*1985*	*1990*	*1995*
Mine output[a]	1.56	1.28	1.18	1.10	1.59	1.89
Output share[b]	30	22	20	17	22	23
Net imports[c]	0.12	0.06	0.39	0.19	–0.07	0.08
Costs[d]	1.04	1.02	0.90	0.70	0.65	0.61
Price[e]	1.83	1.43	1.57	0.80	1.23	1.19

[a] Mine output is measured in millions of metric tons of contained copper. Copper mine output was depressed by an unusually severe economic recession in 1975 and by an industry strike in 1980.

[b] Output share is the ratio of U.S. to western world copper mine output multiplied by 100. The western world includes all countries except those currently or formerly with centrally planned economies.

[c] Net imports reflect U.S. imports of copper minus U.S. exports of copper contained in ores, concentrates, blister, and refined metal, measured in millions of metric tons.

[d] Costs are the weighted-average, adjusted, break-even costs for U.S. producers, measured in real (1990) U.S. dollars per pound. Break-even costs are the costs of producing copper to refined metal minus capital costs (specifically, depreciation, amortization, and interest on external debt) minus any credits for co-product and by-product revenues. Adjusted break-even costs are break-even costs minus the difference, if any, between reported revenues and the product of the world copper price times mine output. Thus, adjusted break-even costs reflect the lowest market price for refined copper at which the mine can operate without cash losses. The figure shown for 1995 is actually for 1993. Costs in nominal dollars are converted to real (1990) dollars using the U.S. GDP implicit price deflator.

[e] Price is the average U.S. domestic producer price, measured in real (1990) U.S. dollars per pound. Prices in nominal dollars are converted to real (1990) dollars using the U.S. GDP implicit price deflator.

Sources: Mine output, output share, net imports, and prices: *Metal Statistics* (Frankfurt am Main: Metallgesellschaft AG, annual); U.S. Bureau of Census; U.S. Bureau of Mines. Costs: Rio Tinto Mine Information System.

A number of companies, including Amoco Minerals, Arco/Anaconda, Cities Service, and Louisiana Land and Exploration, left the industry. Their copper properties were sold to other firms, spun off as independent companies, or simply shut down.

Within the U.S. industry, however, a handful of companies refused to quit. Aware they could not influence the market price, their managers nevertheless believed that copper mining in the United States could once again be made profitable by controlling costs. The decisions these companies took during the bleak years of the early 1980s produced one of the most dramatic turnarounds in mining history.

By 1995, copper mine output had jumped to 1.89 million tons of contained copper, an amount 21% above its 1970 level and 72% above its 1985 level. The U.S. share of western world production at 23% was back to its 1975 level, though still below its 1970 level, as copper demand and western world copper production continued to grow over the 1970–1995 period. More important for producers, net imports had fallen to only 4% of domestic demand, and the gap between average break-even costs and the market price for copper had widened sufficiently to make domestic mining once again profitable and new investment in this industry attractive.

While the U.S. industry as a whole was enjoying a renaissance, recovery at the individual mine level was mixed. Table 4-2, which shows the sig-

Table 4-2. U.S. Copper Output in Thousands of Metric Tons of Contained Metal by Mine, 1975, 1985, and 1995.

Mine	*1975*	*1985*	*1995*
Morenci	121.9	243.1	403.5
Bingham Canyon	167.8	30.6	307.5
Chino	51.2	108.5	156.9
Ray	46.9	80.9	149.9
San Manuel	88.1	88.9	116.9
Sierrita	82.2	97.8	112.4
Mission	103.0	53.4	101.8
Bagdad	17.8	78.4	97.2
Pinto Valley	60.5	78.7	88.7
Tyrone	70.8	135.8	67.9
Cyprus Miami	42.9	69.9	58.5
Butte	88.1	0.0	51.2
Flambeau	0.0	0.0	39.3
White Pine	68.7	1.1	33.9
Superior	38.0	0.0	18.4
Cyprus Tohono	0.0	5.9	15.4
Miami East	10.6	3.7	10.5
Continental	14.2	0.0	7.4
Yerington	30.8	0.0	6.5
Silver Bell	17.8	4.2	3.2
Mineral Park	16.5	1.7	1.4
Ajo	31.0	0.0	0.0
Battle Mountain	14.1	1.4	0.0
Bisbee	10.5	1.6	0.0
Esperanza	12.8	4.5	0.0
Ruth McGill	27.1	0.0	0.0
Sacaton	19.9	0.0	0.0
Twin Buttes	13.8	15.2	0.0
All other mines	51.0	36.7	25.6
Total	1,318.0	1,142.0	1,874.0

Source: Brook Hunt and Associates Limited.

nificant copper mines in the United States, along with their output in 1975, 1985, and 1995, reflects several notable characteristics of the recovery.

First, it was not ubiquitous. Of the twenty-six significant mines producing copper in 1975, seven were shut down by 1995. Another seven had sharply curtailed their output.

Second, only two mines—Flambeau and Cyprus Tohono—producing copper in 1995 were opened after 1970. Together, they accounted for 3% of U.S. output in 1995.[2] The rest came from mines that had been in operation for twenty years or more.

Third, the recovery of the U.S. copper industry was largely the result of expanding output at a few mines—Morenci, Bingham Canyon, Chino, Ray, San Manuel, Sierrita, and Bagdad. Five companies—Phelps Dodge, Kennecott, Asarco, Magma Copper, and Cyprus Amax—owned these mines. These are the companies that refused to abandon what many considered a dead or dying industry.

BEHIND THE REVIVAL

How did this handful of companies manage during the 1980s and 1990s to reestablish the global competitiveness of copper mining in the United States? This section examines five possible explanations: an increase in the production costs of foreign producers, a surge in copper prices, a rise in by-product revenues, a decline in the real wages of domestic copper workers, and finally an increase in U.S. labor productivity.

Higher Production Costs Abroad

Over the longer run, the costs of producing copper abroad have fallen, and fallen substantially. Figure 4-1 shows adjusted break-even costs in real 1990 dollars over the 1972–1993 period for western world copper producers as a whole (including the United States) and for Chile. The latter is currently the largest copper mining country, accounting for about 30% of western world output, and the most important competitor of the U.S. industry. Adjusted break-even costs for the western world fell from 87 to 58 cents between 1972 and 1993, a decline of 33%. The drop in production costs was even more spectacular for Chile. The U.S. industry, thus, has had to reduce its costs substantially over the past couple of decades just to keep pace with producers in the rest of the world.

Since the mid-1980s, however, production costs abroad have risen somewhat, reversing at least for a time the longer-term downward trend. Western world break-even costs, for example, rose from 47 to 58 cents per

Figure 4-1. Average Adjusted Break-Even Costs for the Western World and for Chile in Real 1990 Dollars, 1972–1993.

Source: Rio Tinto Mine Information System.

pound between 1986 and 1993. This helped domestic producers compete both at home and abroad and contributed to the revival of U.S. copper industry.

Some of the factors behind the recent rise in production costs abroad—and the long-run decline—are easy to identify. The sharp appreciation of the dollar over the 1979–1985 period, in large part the result of macroeconomic policies that raised domestic interest rates to curb inflation, lowered production costs abroad when expressed in dollars. Subsequently, however, the depreciation of the dollar, which by 1988 had returned the dollar exchange rate to roughly its late-1970s level, helped the U.S. industry by increasing the costs expressed in dollars of foreign producers. In addition, there is now considerable evidence that state mining companies in Zambia, Zaire, and other countries by the mid-1980s were encountering higher costs as a result of their failure in earlier years to invest sufficiently in development and maintenance (Chundu and Tilton 1994).

Surge in Copper Prices

While copper prices over the longer run presumably follow production costs,[3] in the short run they, like the prices of other metals, are known for their volatility. A surge in demand caused by a boom in the business cycle or an interruption in supply caused by a mine closure can cause the market price of copper to rise and for some time to remain significantly above its long-run trend. Similarly, during recessions and other periods of excess supply, the price may languish far below the full production costs of many producers.

This raises the possibility that cyclical fluctuations in copper prices contributed to the decline of the U.S. copper industry during the 1970s and early 1980s, and then to its subsequent recovery. Figure 4-2 provides some support for this. It shows the London Metal Exchange price of copper in real 1990 dollars over the 1970–1995 period, along with its logarithmic trend. The long-run trend in prices, like that for production costs, is downward. Actual prices, however, are below the trend for many years between 1970 and 1985, while the opposite is the case for several years after 1987.

The sharp rise in real prices in the late 1980s, which was largely unexpected, provided a boost to U.S. and foreign producers alike. It reflected a tightening of supply and demand caused in part by unusually strong growth in the consumption of copper in Asia and the United States. The closure in 1989 of the Bougainville mine in Papua New Guinea, the result of a local insurrection, along with the collapse of production in Zambia and Zaire simultaneously, constrained the available supply.

Greater By-Product Revenues

Where gold, silver, molybdenum, and other by-products are recovered along with copper, they reduce the adjusted break-even costs of producing copper. Such by-product credits, which can vary greatly from year to year due to the volatility of metal prices noted earlier, may help explain the changing fortunes of the U.S. copper industry over the past quarter century.

Figure 4-3 shows the average by-product revenues realized by U.S. copper producers over the 1970–1993 period as a percentage of their cash

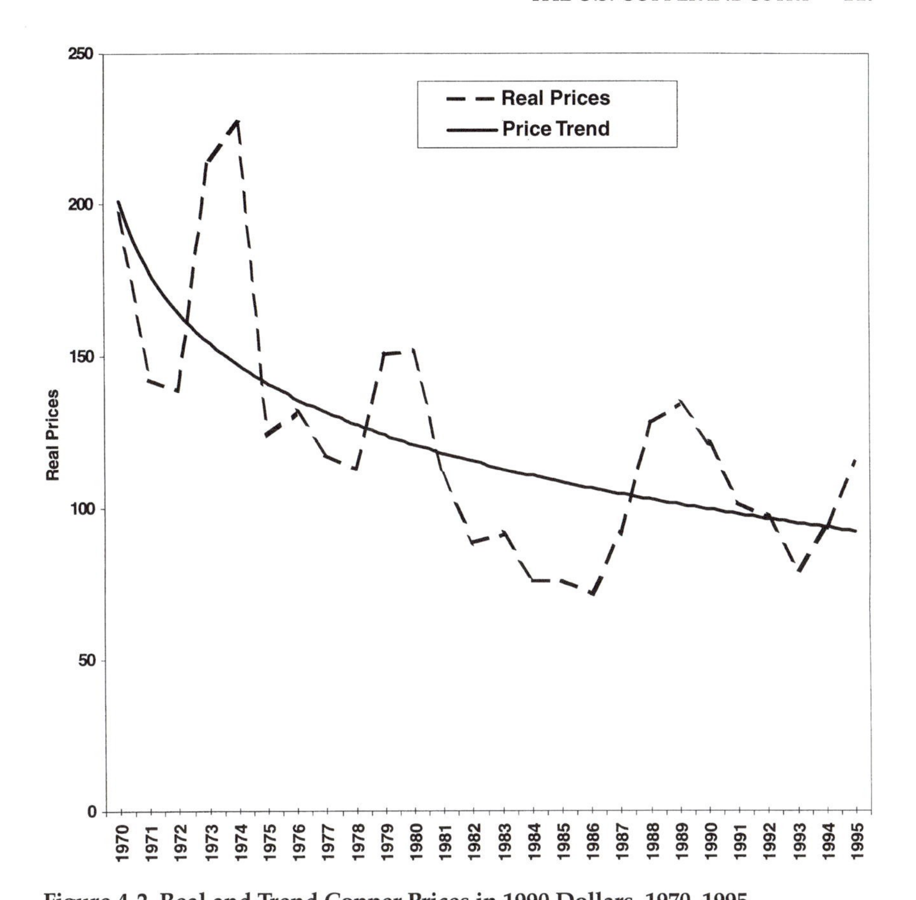

Figure 4-2. Real and Trend Copper Prices in 1990 Dollars, 1970–1995.

Note: The price shown is for high grade and grade A copper on the London Metal Exchange.

Source: Commodity Exchange of New York, *Metals Week*, and *American Metal Market* as reported by the U.S. Geological Survey.

costs. (Cash costs are break-even costs plus by-product revenues.) This figure indicates that by-product revenues covered about the same portion—roughly one-fifth—of the cash costs of producing copper in the United States in the early 1970s as in the early 1990s. So changing by-product revenues have not affected the competitiveness of U.S. producers over the long run.

Between the early 1970s and early 1990s, however, by-product revenues followed two pronounced cycles. The percent of by-product revenues to cash costs rose sharply during the 1970s, reaching nearly 50% by 1980. This favorable trend offset to a considerable degree the rise in real

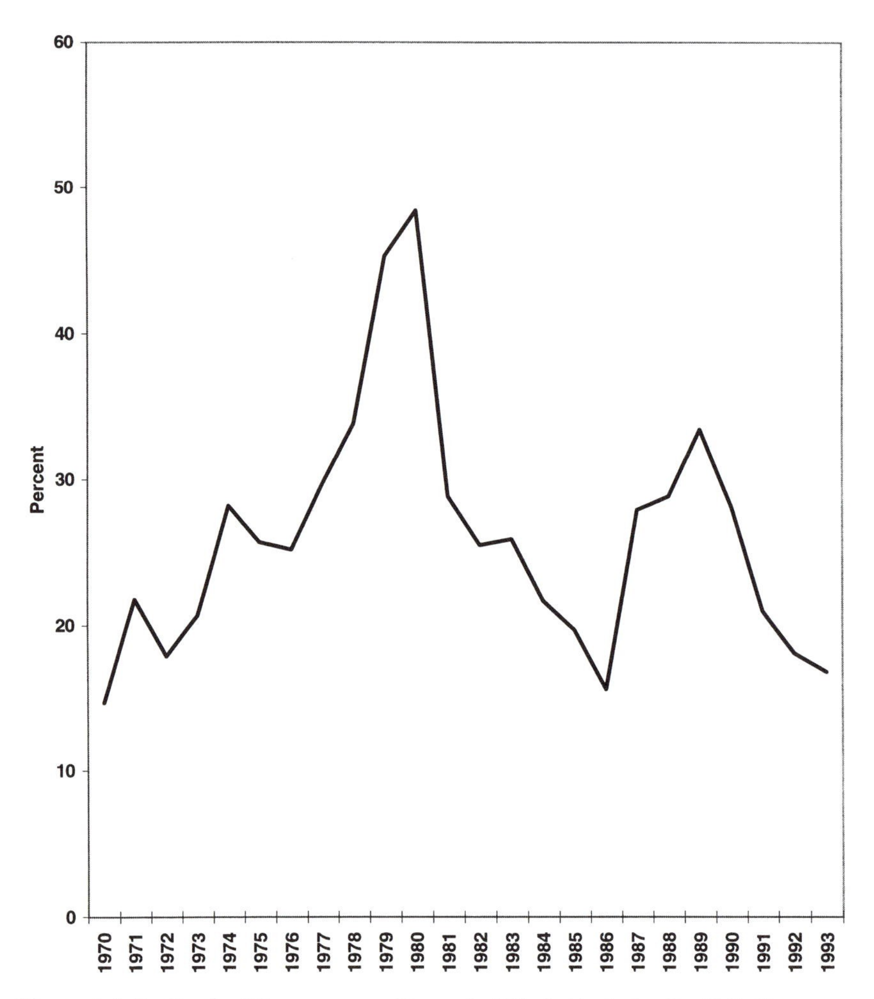

Figure 4-3. By-Product Revenue as a Percent of Cash Costs for the U.S. Copper Industry, 1970–1993.

Source: Rio Tinto Mine Information System.

wages and other costs over these years. It was not to last, however. During the early 1980s, the percentage of cash costs covered by by-product credits dropped as rapidly as it had risen in the late 1970s, contributing in the process to the industry's troubles during this period.

The second cycle was more modest. The percentage of cash costs covered by by-product credits began to rise in 1987 and for several years contributed to the industry's revival. The decline set in about 1990. Part of this decline can be attributed to the growing importance of a new production process, solvent extraction–electrowinning, which is discussed more fully below. This process does not recover any by-products.

Lower Wages

Trends in real wages also affected the changing fortunes of the U.S. copper industry over the past several decades. As Figure 4-4 illustrates, real hourly wages in copper mining and milling, after rising persistently for more than three decades, plummeted by more than 25% between 1984

Figure 4-4. U.S. Real Hourly Wages and Labor Input per Unit of Output for Copper Mining and Milling, 1950–1994.

Source: U.S. Bureau of Labor Statistics.

and 1989. They have since remained fairly constant at levels similar to those of the early 1970s.

This sharp reversal in the long-run upward trend in real wages was not easily achieved. Phelps Dodge confronted organized labor directly and suffered a long and bitter strike during 1983 at its properties in Arizona. It continued to produce during the strike, and ultimately union members who resisted the elimination of the cost-of-living adjustment and other concessions were permanently replaced.

Kennecott shut down the Bingham Canyon mine in early 1985 after five years of consecutive losses and started a $400 million modernization program at this mine later the same year. The union agreed to a new contract in 1986 that gave the company much greater flexibility in work rules and staffing assignments as well as an average 25% cut in salary and benefits. According to the president of the United Steelworkers of America Local 392 at the time (Hamilton 1988, 1):

> We paid dearly to come back to work, but we had to do it. Changes had to be made.... I think we did the right thing, but a lot of people think we sold out to Kennecott.

Some companies, including Magma Copper and Inspiration Consolidated Copper, offered profit-sharing plans tied to the price of copper to help convince workers to accept a reduction in salary and other concessions.

These arrangements produced sizable bonuses in the later 1980s when copper prices rose sharply (Regan 1988, 1). This raised the possibility that a relapse in prices and profits might adversely affect worker morale and productivity. Magma, as a result, shifted in 1989 from a profit-sharing plan to what it called a gain-sharing plan. The latter tied bonuses directly to productivity and cost performance, rather than to profits and thus the price of copper, over which managers and workers have no control.

At the White Pine Mine in northern Michigan, the union's hard line against any concessions led to a strike in 1983. The mine remained shut for two years. During this period, Louisiana Land and Exploration sold the mine along with the other assets of the Copper Range Company to Echo Bay Mines, which in turn resold the mine to the mine's employees led by a former president of the Copper Range Company. In return for a share of the company and perhaps more importantly an opportunity to return to work, workers accepted a cut in pay of nearly $4.00 per hour, or about 33% (McDaniel 1989). In addition, the cost-of-living adjustment was dropped. While these conditions caused some consternation at the time, most workers realized the old pay schedules were no longer feasible. Moreover, four years later when the mine was sold to Metallgesellschaft, the large German mining company, the average worker received $60,000 for his ownership shares (Stertz 1989, 1).

Increased Labor Productivity

While the fall in real wages, the increase in by-product revenues, the jump in copper prices, and the rise in production costs abroad during the 1980s played a part in the recovery of the U.S. copper industry, the surge in labor productivity between 1980 and 1986 was particularly important. The hours of labor required to mine and mill a ton of copper, as Figure 4-4 shows, fell by more than 50% during this period. This means one worker in 1986 was producing copper at a rate equivalent to that of two workers just six years earlier.

This rise in labor productivity is part of a longer-run trend that has dramatically reduced the hours of labor to mine and mill a ton of copper since 1950. The long-run trend, however, follows a series of downward steps, rather than a continuously declining curve, with major increases in productivity occurring in the 1950s and then, as just noted, in the early 1980s. In the 1960s and 1970s, labor productivity changed little. The same is true for the decade since 1985.

FACTORS AFFECTING LABOR PRODUCTIVITY

The critical role played by rising labor productivity in the revival of the U.S. copper mining industry leads naturally to a search for possible causes—increases in capital and other inputs available per worker, changes in the quality of the copper ore being mined, and new technology and other innovative activities.

Capital and Other Inputs per Worker

During the 1980s, Kennecott increased labor productivity at its Bingham Canyon mine by nearly 400%. The $400 million modernization program, mentioned earlier, that the company began in 1985 is credited with much of this increase (Goldberg 1991; Carter 1990). This raises the question: how much of the increase in labor productivity in copper mining in the United States over the past several decades is simply the result of workers' having more capital and other inputs with which to work?

Efforts to answer this question, and in turn to measure trends in total or multifactor productivity, are hampered by the absence of data on the capital stock and intermediate goods used in copper mining in the United

States. The only attempt of which we are aware to estimate nonlabor inputs and multifactor productivity for copper mining in the United States is Parry's chapter in this book. By making a number of assumptions, Parry derives estimates of the capital stock (structures, equipment, inventories, and land) and intermediate inputs (energy, purchased services, and raw materials) for copper on the basis of data available for the U.S. metal mining industry as a whole.

The findings, which Table 4-3 highlights, are quite interesting. Between 1972 (the first year reported by Parry) and 1980, copper mine output fell by 22% in the United States. Hours of labor fell by nearly the same percentage, leaving labor productivity little changed over this period. In contrast, the estimated capital stock rose by 10% and inputs of intermediate goods by about 12%. As a result, multifactor productivity in copper mining declined by 22%.

Table 4-3 also reflects the recovery of U.S. production during the 1980s and early 1990s. The jump in output was achieved with substantially smaller inputs of labor, causing labor productivity to nearly triple between 1980 and 1992. Multifactor productivity also rose, though less dramatically, as inputs of labor, intermediate goods, and capital per unit of output all declined.

The faster growth in labor productivity over multifactor productivity implies that some of the increase in labor productivity came about because labor had more capital and intermediate goods with which to work. While this finding is consistent with casual observation and anecdotal evidence, two caveats should be noted.

First, the numbers in Table 4-3, as pointed out earlier, are proxies based on data for all metal mining. While they represent the best estimates available of multifactor productivity, they rest on a number of critical assumptions and in particular do not take account of the writing-off of capital that occurred as many copper mines were closed prematurely.

Table 4-3. Indices of Production, Labor Input, Capital Stock, Intermediate Goods Input, Labor Productivity, and Multifactor Productivity for the U.S. Copper Mining Industry, Selected Years (1972 = 100).

	1972	*1975*	*1980*	*1985*	*1990*	*1992*
Production	100	84	78	73	105	116
Labor input	100	104	80	35	41	41
Capital stock	100	114	110	95	79	81
Intermediate goods input	100	99	112	65	130	119
Labor productivity	100	81	97	208	257	284
Multifactor productivity	100	81	78	104	103	123

Source: Chapter 6 of this book, Figures 6-1, 6-3, 6-5, and 6-6.

Second, while Table 4-3 reflects an increase in the quantity of capital and intermediate inputs per unit of labor, the most dramatic trend it highlights is the drop in all inputs per unit of output between 1980 and 1992. This suggests that new technologies and other innovative activities have had a major impact on both labor and multifactor productivity since 1980.

Moreover, investments in structures and equipment often embody new technologies and other innovative developments. Thus, in practice, the effects of an increase in capital on productivity cannot easily be separated from those of new technology. The $400 million that Kennecott invested in the Bingham Canyon mine during the 1980s, for example, paid for a mobile in-pit crusher, a five-mile-long coarse-ore conveyor system, a seventeen-mile slurry pipeline that transports the concentrate from the mill to the smelter, three autogenous grinding mills, six ball mills, and ninety-seven large-capacity flotation machines—all of which embodied state-of-the-art technology. Thus, some of the effect of capital deepening on labor productivity reflects the influence of better technology as well.

Quality of Copper Deposits

One additional factor of production still to be considered is the quality of the mineral reserves being exploited. Companies interested in maximizing shareholder value, it is widely presumed, will tend to mine the highest-quality (and hence lowest-cost) ores first. Thus, over time the remaining reserves will decline in quality, causing productivity to fall, unless other developments offset this adverse effect.[4]

This tendency toward poorer reserves at individual mines, however, does not necessarily mean the average reserve quality across all mines has to fall. For the country as a whole, the average can rise (a) as a result of developing new mines or expanding output at existing mines with above average reserves or (b) as a result of closing down or cutting back production at mines with poor reserves. Thus, the jump in productivity since 1980 could in part reflect the mining of higher-quality ores.

We already know that the discovery and development of new mines has been of little or no importance. As pointed out earlier, Flambeau and Cyprus Tohono were the only two significant mines producing copper in 1995 that were not in operation in 1975, and together they accounted for but 3% of U.S. mine output.

This implies that the quality of U.S. copper reserves either declined over time or at best remained largely unchanged. Still, the quality of the ores actually being mined may have increased as production shifted among mines. As noted above (and shown in Table 4-2), the United States had twenty-six significant copper mines in 1975. Seven of these mines were shut down and production at another seven sharply curtailed over the following two decades. Many of the remaining mines increased their output, some significantly, over this period. If the latter possess the better ores, which seems likely, this shift may have more than offset the decline in reserve quality at the level of individual mines.

Figure 4-5 allows us to assess this possibility. It shows what is commonly called head grades—the average grade of the copper ore actually mined—at all mines and at open-pit mines in the United States over the

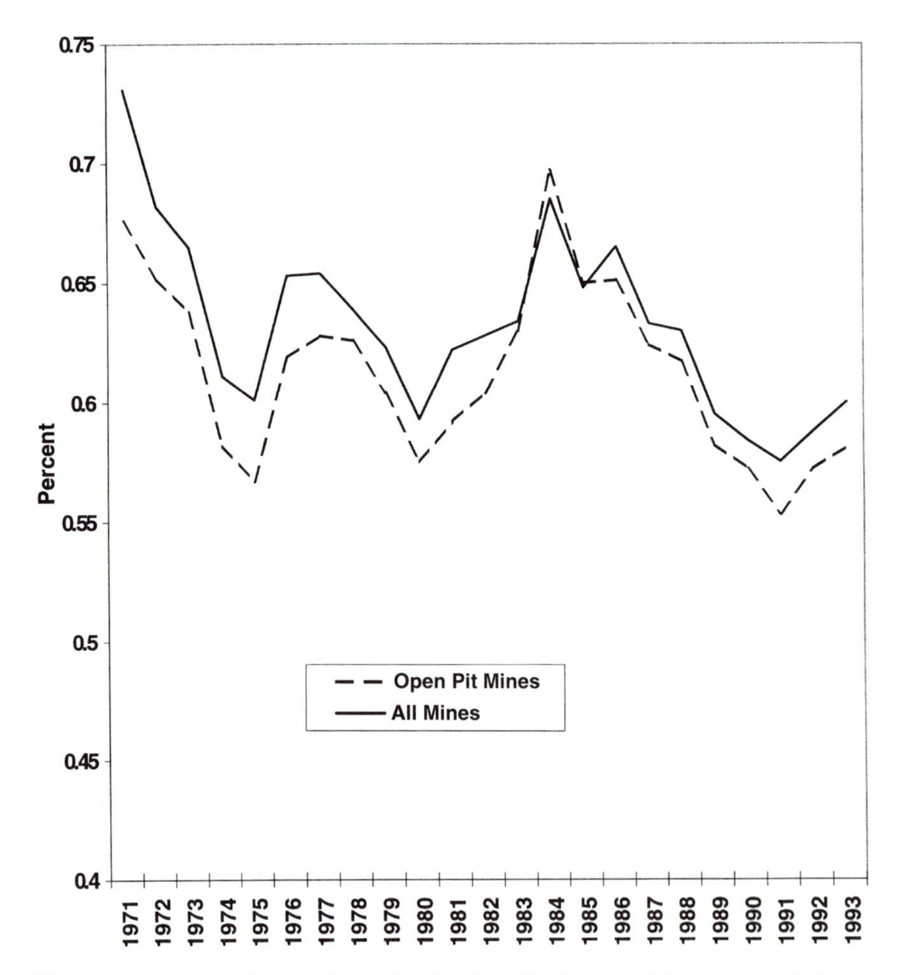

Figure 4-5. Average Copper Head Grades for All Mines and for Open-Pit Mines in the United States, 1971–1993.

Source: Rio Tinto Mine Information System.

1971–1993 period. While grade alone does not determine reserve quality, it is a major consideration.[5]

The average head grade for all types of copper mining declined from 0.73 to 0.60% between 1971 and 1993, suggesting that the quality of the reserves mined in the United States has fallen, not risen, over the past couple of decades. The decline, however, was not monotonic. Average head grade rose in 1976–77 and during the early 1980s when copper prices were depressed. The second increase contributed to the jump in labor productivity and the revival of the U.S. copper industry during the 1980–1986 period. This stimulus, however, was short lived, as average head grade fell during the second half of the 1980s, reaching a historic low by the early 1990s.[6]

Thus, changes in the grade of ore mined—the rise during the early 1980s and the decline during the late 1980s—help explain the sharp jump in labor productivity during the first half of this decade and the stagnation in labor production thereafter. Over the entire 1971–1993 period, however, changes in head grades, since head grades have fallen, do not explain the dramatic rise in labor and multifactor productivity.

Innovative Activity and New Technology

As noted earlier, significant declines in the labor, intermediate goods, and capital required to mine and process a ton of copper after 1980 (Table 4-3) suggest that innovative activity and new technology contributed greatly to the growth in labor productivity and the recovery of the U.S. copper industry. A number of studies (Office of Technology Assessment 1988; National Materials Advisory Board 1990; Queneau 1985) provide further support for this proposition by documenting the changes across the entire spectrum from exploration to recycling that have taken place in the way copper and other metals are produced and consumed.

One particularly important development over the past several decades has been the increasing use of the solvent extraction–electrowinning (SX-EW) process. The big advantage of this new technology lies in its low operating and capital costs. Pincock, Allen, and Holt (1996), on the basis of its review of SX-EW facilities in the United States and abroad, estimated the average operating costs of producing SX-EW copper in 1995 at 39 cents per pound—8 cents for mining and handling, 13 cents for leaching, 16 cents for SX-EW recovery, and 2 cents for general and administrative costs. This total compares quite favorably with the 60 cents per pound the study estimated as the average operating costs for copper produced with the traditional technology. Capital costs of building new SX-EW capacity—estimated at $3,400 per annual ton of additional capacity at greenfield plants and $1,700 at existing plants—also are considerably lower than those for traditional facilities. U.S. companies, as the next section shows, helped pioneer the development of this new technology and led the world in its adoption. Between 1980 and 1995, the share of total U.S. copper output accounted for by the SX-EW process increased from 6 to 27%. By the end of this period, the country possessed more than half of the world's SX-EW capacity.

SX-EW technology is just one of many innovative activities that U.S. producers undertook to increase labor productivity and reduce costs. Porter and Thomas (1988) estimated that new mine plans led to a drop in the average stripping ratio, measured as the tons of waste removed per ton of ore mined, from 2.11 in 1981 to 1.33 in 1986.

Other innovations came about as a result of the increased flexibility in work rules and manning assignments that new agreements with labor made possible. At the Morenci and Bingham Canyon mines, the ore-handling system was improved. Large trucks and in-pit mobile crushers with conveyor belts replaced rail haulage, once favored in very large mines for its ability to handle huge tonnages. The computerization of truck scheduling and real-time process controls in mills produced significant economies at some sites. New work schedules helped at other operations. Larger trucks, shovels, and drills coupled with more cost-effective explosives also generated savings. The concentration of U.S. production at a smaller number of large mines produced savings from economies of scale.

Other Factors

To explain the surge in labor productivity during the 1980s, which in turn played such a critical role in the recovery of the U.S. copper industry, we have focused on capital and other inputs available per worker, head grades, and innovation. While all three are to some extent important, innovation and new technology are particularly so.

Other factors, which we have not considered, also affected productivity. The intriguing role of management in the industry's decline and recovery is a story of its own and beyond the scope of this inquiry.

More stringent government regulations in the areas of the environment and of worker health and safety also have had an impact. As much as they may be desirable for other reasons, they presumably have inhibited, rather than fostered, productivity growth and thus provide little insight into the recovery of the U.S. copper industry.[7] What they do suggest, which is important and worth noting, is that the industry was not saved by relaxing environmental regulations and allowing it to increase its pollution. To the contrary, copper producers over this period were meeting increasingly demanding and costly standards.

SOLVENT EXTRACTION–ELECTROWINNING

While innovative activity and new technology clearly can and do increase labor and multifactor productivity, can they reduce any particular company's or country's comparative costs, that is, its costs relative to its competitors' costs? Skeptics point out that new innovations, such as larger trucks and better drills, are often produced by equipment makers who are not only willing to sell their products abroad but actually promote such sales. For this and other reasons, new innovations diffuse very rapidly, particularly in what is becoming an increasingly global economy, and any cost advantage a firm realizes from introducing a new technology will be short-lived at best.

The validity of this view, however, depends on two implicit assumptions. First, new innovations are one-time, discrete events, rather than an ongoing series of minor and major advances. Where development is continuous over an extended period of time, innovators can maintain a technological lead and in turn a cost advantage. Second, new innovations are neutral in the sense that they possess the potential to reduce all producers' costs similarly. This need not be the case. Large trucks, for example, are particularly cost-effective in large open-pit mines and simply unusable in small underground operations.

While these two conditions may be satisfied for some innovations, anecdotal evidence indicates that they frequently do not hold and thus that innovative activity often shifts the comparative costs of producers, favoring some at the expense of others. To illustrate this point, we will delve in some depth into the early development and subsequent evolution of the SX-EW process. Before doing so, however, we examine the nature of the SX-EW process and compare it with the traditional technology for processing copper ores.

Copper Technology

The traditional technology entails mining sulfide copper ore in underground or open-pit mines. The ore is then moved by truck, rail, or conveyor belt to a mill where it is crushed and the copper-bearing minerals are separated from the waste material or *gangue* by flotation. The resulting concentrate (25–40% copper) is shipped to a smelter for partial purification (97–99% copper) and then on to a refinery for electrolytic purification (99.99% copper).

The SX-EW process involves first leaching existing mine dumps, prepared ore heaps, or in situ ore with a weak acidic solution. The solution is recovered and in the next stage—the *solvent extraction* stage (SX)—mixed with an organic solvent (referred to as an *extractant*), which selectively removes the copper. The copper-loaded extractant is then mixed with an aqueous acid solution, which strips it of the copper. The resulting electrolyte is highly concentrated and relatively pure and is processed into high-quality copper in the third and final stage by *electrowinning* (EW).

The leaching of mine dumps from past operations does not require mining since these dumps are already in place. Mining is similarly not necessary for in situ leaching, where copper is extracted from fractured ore remaining in place within the original deposit. The leaching of prepared ore heaps, however, does require that the ore be removed from the deposit, in some cases crushed, and then placed on prepared heaps or "lifts." Thus, the SX-EW process often, though not always, avoids the substantial costs of mining.

SX-EW is a hydrometallurgical process, while the traditional technology is referred to as a pyrometallurgical process since it entails smelting. There are, however, other hydrometallurgical processes for treating copper ores. For centuries, copper has been leached and then recovered by precipitation with iron. This process produces cement copper (85–90% copper), which then must be smelted and electrorefined. Direct electrowinning involves leaching and electrowinning, but skips the solvent extraction stage. It produces a lower-quality copper contaminated with iron and zinc, which requires further processing. As a result, the product of the direct electrowinning process is considered more a substitute for high-grade (number 1) copper scrap than for electrorefined copper.

Evolution of SX-EW Technology[8]

The leaching of copper ores can be traced back to fifteenth century Hungary, and, until earlier in the twentieth century, leaching and precipitation by iron was the only hydrometallurgical method available to process copper. With the commercial development of electricity, the direct electrowinning of copper became possible. The first commercial application of this technology took place in 1915 at Ajo in Arizona and at Chuquicamata in Chile. As noted earlier, neither of these processes produced copper of sufficiently high quality to compete directly with electrorefined copper. This became possible only with the development of solvent extraction technology during and after World War II.

Solvent extraction was first employed during the war to separate and recover uranium. It is now widely used to treat a variety of metals—copper, zinc, nickel, gold, silver, rare earth metals, and others. It typically concentrates the metal in solution and acts as a filter to remove impurities.

Ranchers Exploration and Development Company began operating the first commercial SX-EW facility for copper at its Bluebird Mine in Arizona in March 1968. It worked closely on this project with the General

Mills Corporation (whose chemical division developed the necessary extractants during the early 1960s), with Hazen Research Incorporated (which as a consulting engineering firm helped develop the process), and with the Bechtel Corporation (which designed and constructed the production plant).

The second SX-EW facility started in 1970 at the Bagdad mine of Cyprus Mines. Subsequent installations occurred during the 1970s at Nchanga in Zambia, at Cerro Verde in Peru, and at Johnson Camp, Twin Buttes, Miami, Battle Mountain, Inspiration, and Ray in the United States.

By the early 1980s, SX-EW was rapidly replacing direct electrowinning and leaching with iron precipitation. The quality of SX-EW copper did not yet rival that of electrorefined copper. It was generally sold at a discount and used by alloy producers and other customers who did not need the highest-quality copper.

Shell, Acorga Limited, and Ashland Chemicals joined General Mills as suppliers of extractants. Henkel Corporation purchased the operations of General Mills and Shell, and along with Acorga it remains a major supplier of extractants. These companies developed second- and third-generation extractants that enhanced the removal of impurities. The new extractants also extended the range of treatable leach solutions, in particular allowing the upgrading and purification of much more acidic solutions with a wider range of copper concentrations. This greatly increased the dump and heap materials from which copper could be processed. At the same time, improvements in equipment design permitted the more rapid and effective mixing of the extractant with the leach solution and then the recovery or stripping of the copper from the extractant.

The successful introduction of SX-EW stimulated a search for better technologies at the leaching and electrowinning stages of production. For example, although electrowinning and electrorefining are similar, electrowinning uses a lead rather than copper anode. Early problems with lead contamination were eventually overcome by the addition of cobalt salts to the electrolyte.

At the leaching stage, important advances have occurred in the acid solutions or lixiviants used to dissolve the copper in a continuing effort to extend the range of copper minerals suitable for SX-EW processing. Copper is recovered from both primary and secondary copper-bearing minerals. The former are the primary sulfides (chalcopyrite, bornite). The secondary minerals, produced by natural processes over extended periods of time from the primary minerals, include carbonates (azurite, malachite), oxides (cuprite, tenorite), hydroxy-silicates (chrysocolla), sulfates (antlerite, brochantite), native copper (copper metal), and secondary sulfides (chalcocite, covellite). The nonsulfide ores are typically the easiest and quickest to leach. The sulfide ores, both primary and secondary, require that an oxidant be added to the acidic leach solution. The process takes longer, for some ores literally years longer, and is less complete. Bioleaching, which uses bacteria as a catalyst, has shortened the leaching time for sulfide ores. Even though most of the world's copper reserves are primary sulfide minerals and thus still best processed with the traditional pyrometallurgical technology, these efforts have expanded, and continue to expand, the use of SX-EW.

Other advances include the recent addition of surfactants to the leach solution, which allows the solution to spread more completely over the copper-bearing minerals, increasing the amount of copper recovered. For some time, efforts have also been under way to improve in situ leaching techniques, which eliminate the need and cost of mining. In arid regions, producers have replaced sprinklers with drip irrigation and covered their heaps with plastic sheeting to reduce solution losses from evaporation.

Technological advances in SX-EW technology over the past several decades have reduced production costs at all three stages of production—leaching, solvent extraction, and electrowinning. They have also extended the range of ore sources from which copper could be economically extracted. In the 1970s and early 1980s, SX-EW was primarily used to recover copper from existing waste piles, which required no mining and contained copper minerals that were quick and easy to leach. In recent years, the process has been increasingly used in mine-for-leach operations, where certain types of copper minerals are mined for leaching. Though not yet successful, efforts continue to develop techniques to use SX-EW in place of the traditional pyrometallurgical technology to process the primary sulfide minerals that account for the bulk of the world's copper reserves.

This review, though it identifies only some of the important advances in the SX-EW process over the past several decades, illustrates the dynamic nature of this technology. With such technologies, innovative firms can maintain a technological lead even though competitors adopt any particular advance within a few years or even months of its introduction. Maintaining such a lead, however, requires a persistent commitment to innovation and to the development and use of new technologies.

Effects on Comparative Costs

The SX-EW process has in two ways helped the U.S. copper producers reduce their costs relative to those of their foreign competitors. First, the technological lead in this field that U.S. firms have enjoyed since the introduction of the solvent extraction process in the late 1960s has allowed them to benefit from the cost savings derived from new advances in SX-EW technology sooner than other producers. If the U.S.

producers maintain their technological lead and continue to pioneer important new advances in the SX-EW process, this advantage can continue indefinitely.

Second, the SX-EW process is not a neutral innovation. It reduces the costs of some copper producers much more than others. In particular, it tends to favor:

- countries, such as the United States, that have historically been important copper producers and that over the years have accumulated substantial waste piles of oxide copper minerals. These minerals are often found near the surface of porphyry deposits. As the traditional pyrometallurgical technology cannot process such ores, they have been removed and placed into waste dumps. SX-EW technology now can treat these dumps quite inexpensively. In addition, the SX-EW process allows the recovery of copper from low-grade ores that producers in established mining districts have not found economic to extract with the traditional technology.
- countries, such as the United States, that are currently important producers of copper and that enforce stringent environmental regulations. Producers in these countries are forced to recover the sulfur emissions from their smelting operations, typically in the form of sulfuric acid. This provides a low-cost source of leach solution, as diluted sulfuric acid is the prevailing acid used at this stage of the SX-EW process. There is, thus, a strong symbiotic relationship between the new SX-EW process and traditional technology.
- countries, such as the United States, whose copper deposits are located largely in arid regions. Where precipitation is heavy, maintaining the desired characteristics of the leach solution is difficult.
- countries, such as the United States, that possess substantial copper reserves with few or no by-products of value. One current shortcoming of the SX-EW process is its inability to recover gold, molybdenum, and other valuable by-products. The SX-EW process also is not suitable for treating primary sulfide minerals, such as chalcopyrite, given the lengthy periods required to leach these minerals.

The inherent advantages that the SX-EW process bestows on the United States are not accidental. U.S. producers pushed the development of this new technology because of the potential benefits it offered for their operations. While future developments may strengthen or weaken these advantages, as long as U.S. producers remain the major innovators they will have strong incentives to pursue most actively those developments that particularly favor their activities.

Others, of course, may benefit as well. Indeed, the conditions that make SX-EW advantageous to U.S. producers exist in other countries.

This is especially so for Chile, which has recently surpassed the United States in the production of SX-EW copper. The important point, however, is that SX-EW has reduced the costs of some producers more than others. The United States has benefited from this nonneutral effect on comparative costs and may continue to do so for some time.

LESSONS

The story of the U.S. copper industry over the past three decades—its decline and dramatic revival—is of some intrinsic interest, particularly for those whose welfare is significantly affected by the fortunes of this particular industry. We, however, have studied this industry in the hope of finding implications and lessons that extend beyond copper mining in the United States. This final section first looks at the implications for mining and mineral-producing countries competing in increasingly competitive global markets and then for all countries striving to increase the productivity of their work forces.

Comparative Advantage in Global Mineral Markets

Why countries produce and export the goods they do has interested economists for centuries. Modern explanations, based on the doctrine of comparative costs that David Ricardo introduced nearly two hundred years ago, contend that states will produce and export those commodities whose production costs are low relative to the costs of other domestic products when compared with production costs in other countries.

The doctrine of comparative costs is just the first step in accounting for comparative advantage. To be useful, it must be coupled with an explanation of why differences in comparative costs arise among countries. In the first attempt to answer this question, Ricardo and other classical economists pointed to differences in labor productivity.

The factor endowment theory, advanced by the Swedish economists Eli Heckscher and Bertil Ohlin in the early years of the twentieth century, provides a second explanation for differences in comparative costs. This theory contends that differences in production costs are due largely to differences between countries in the price of capital, labor, and other factors of production, which arise because countries enjoy different factor endowments.

Over the past several decades, numerous other explanations for differences in comparative costs have emerged. These theories stress the importance of differences among countries in the generation and diffusion of technology, in human capital, in opportunities to exploit economies of scale, in regional externalities, in domestic market conditions, in strategic trade policies capitalizing on learning economies, and in the opportunities to realize economies of scope.

These theories arose to explain trade between the industrialized countries in manufactured goods and services. While differences in factor endowments and productivity exist among these countries, they are small compared with the differences between the developed and developing countries. Moreover, developed countries with similar domestic conditions, such as those of Western Europe, often import and export the same or similar final goods with each other. This has produced considerable uneasiness over the validity and usefulness of earlier theories in explaining trade between such countries in automobiles, farm machinery, computers, and other manufactured products.

This dissatisfaction, however, does not extend to trade in primary products. Here, one still finds widespread acceptance of the factor endowment theory as the most useful explanation for international differences in production costs. In the words of Haberler (1977, 4):

> The most obvious factors that explain a good deal of international trade are 'natural resources'—land of different quality (including climatic conditions), mineral deposits, etc. No sophisticated theory is required to explain why Kuwait exports oil, Bolivia tin, Brazil coffee and Portugal wine. Because of the deceptive obviousness of many of these cases economists have spent comparatively little time on 'natural resource trade.'

Intuitively, the hypothesis that mineral endowment is the overriding determinant of comparative advantage in mining and mineral processing is very appealing. It also has some important implications.

First, all other determinants of comparative advantage in mining are of second-order importance compared with resource endowment. In particular, the generation and diffusion of new technology along with other innovations are relatively insignificant. This is either because the production process changes little over time or because new technologies and innovations diffuse quickly and effortlessly around the world, providing little opportunity for individual mines, companies, or countries to exploit such developments to achieve a cost advantage.

Second, it suggests that comparative advantage in mining and mineral processing is largely a transitory gift of nature. Countries with the best deposits and the lowest production costs are the most competitive in world markets. Once their deposits are exhausted, comparative advantage shifts to those countries with the next-best deposits. From time to time, new discoveries may also produce shifts in comparative advantage.

Third, and a corollary of the second, corporate managers and labor can do little to maintain or enhance the comparative advantage of a particular mine. As long as it possesses reserves, it will operate. Once its reserves are gone, it will close. Companies can maintain their comparative advantage only by ensuring that new high-quality deposits are discovered or otherwise acquired to replace those being depleted.

Fourth, governments of mineral-producing and -exporting countries are similarly limited in their ability to promote comparative advantage. While they can encourage domestic exploration for new deposits (through favorable land use policies, taxation, permitting requirements, and so on) to slow the decline, eventually the depletion of the best deposits will encourage firms to look abroad for new reserves. Governments can capture and invest some of the rents or profits their domestic reserves generate so that the public welfare can be sustained once the mineral wealth is gone. They cannot, however, prevent mineral exhaustion and the ultimate loss of comparative advantage this effects.

This picture of comparative advantage is largely deterministic. One exogenous factor (mineral endowment) governs the evolution over time of production and trade. Other than finding and developing new high-quality deposits, management, labor, and governments can do little to reduce their relative costs and enhance their comparative advantage.[9]

The interesting policy issues now are how long the endowment will last and how to divide the resulting rents among workers, equity holders, the state as a whole, and other interested parties. These issues lead inevitably to concerns over sustainability, intergenerational equity, and the complexities of green national income accounting.

On the other hand, if extending mineral endowment by developing known marginal deposits or by discovering new deposits is not the overriding determinant of costs, but simply one of many activities mineral producers pursue in an unrelenting struggle to reduce costs, the set of important issues changes. The whole process becomes much more endogenously driven. There are still rents to be captured, but they are no longer predetermined gifts of nature, fixed in size, that firms and countries can effortlessly gather up. They are instead created by mining companies, specifically those companies that succeed in the global competition to reduce production costs.

In this scenario, managers and workers are not helpless bystanders watching external forces unravel their predetermined fate, but instead are crucial players who through their innovative efforts significantly control their own destiny. The role of government shifts from ensuring that society as a whole gets its fair share of the exogenously given rents and that these are used in a manner that ensures intergenerational equity, to providing an economic climate that encourages the innovative activities of firms and individuals.

In this scenario, human ingenuity can keep the real prices of mineral commodities falling indefinitely, making concerns over sustainability and intergenerational equity less pressing. Interestingly, the forces shaping comparative advantage for mineral commodities are not all that different from the forces determining which firms and countries will produce and export high-technology goods and services.

The U.S. copper industry demonstrates that the second scenario, at least in some instances, is more relevant and useful in understanding the nature of production and trade in mineral commodities than the traditional paradigm based on the factor endowment theory. During the 1980s, U.S. producers greatly reduced their production costs, not by finding new copper deposits and improving their resource endowment, but by pursuing a wide variety of innovative activities that substantially raised productivity.[10]

While the dramatic turnaround of the U.S. industry in the 1980s may be exceptional, the experience of the successful U.S. firms in other respects is not all that unusual. Around the world, mineral companies are constantly searching for new technologies and other innovations to reduce costs, knowing that maintaining competitiveness in the future means producing at ever-lower costs. The discovery and development of high-quality deposits are only one of many possible avenues for reducing costs and not always the most important.

Public Policy and Productivity

The story of copper mining in the United States also has implications for countries striving to raise living standards by promoting productivity growth. The U.S. government denied the copper industry's request for protection from imports in 1978 and 1984 on the grounds that the costs to fabricators would exceed the benefits to copper producers. While there is no way to know for certain how copper mining in the United States would have fared with protection, with the benefit of hindsight several conjectures seem reasonable.

Competition from the world's most efficient producers left U.S. firms—their managers and workers—with little choice. They had to either shut down or pursue a range of innovative activities to enhance their productivity and efficiency.

Protection would have limited and constrained the stimulating effects of this competition, weakening the push for new and better production methods. The sharp jump in labor productivity and fall in costs during the 1980s would have come more slowly, perhaps much more slowly. The revival of the industry would have been less spectacular. Copper mining in the United States would now almost certainly be less efficient, less competitive in global markets, and possibly still dependent on government protection or subsidies for its survival. It appears, ironically, that the domestic copper producers were fortunate their requests for protection in 1978 and 1984 went unheeded.

It is important, of course, to remember that there were costs. Mines closed, and some three-quarters of the work force lost jobs. Such costs are the price an economy pays to sustain productivity growth and international competitiveness in dynamic, high-technology industries, including mining. Fortunately, these costs are temporary, while the benefits are long lasting. Still, for societies that care both for those hurt by change and for their children, the challenge is to find ways to help the former without impeding growth in output per worker. For as Paul Krugman so succinctly notes, over time labor productivity and little else matters in determining a country's standard of living.

ACKNOWLEDGMENTS

We are grateful to the William J. Coulter Foundation for their financial support and to Rio Tinto Plc and Brook Hunt and Associates Limited for sharing proprietary data. We would also like to acknowledge the many helpful comments that we received from Michael Bailey, Ross R. Bhappu, David W.J. Coombs, Phillip C.F. Crowson, Daniel L. Edelstein, David Humphreys, Merton J. Peck, Jane Robb, Simon D. Strauss, and Richard Wilson.

ENDNOTES

1. Barnett and Morse, for example, found that an index of real labor and capital costs per unit of output in the minerals sector (where 1929 = 100) fell from 210 at the end of the nineteenth century to 47 by 1957. An index of the per unit labor and capital costs for minerals relative to the per unit labor and capital costs for the nonextractive sector as a whole (that is, all of the economy other than the mineral, forestry, and agricultural sectors) fell over the same period from 154 to 68 (Barnett 1979, Tables 8-1 and 8-2).

2. Cyprus Tohono was shut down in early 1997 due to high production costs. Flambeau was closed in 1997 because of the depletion of its reserves.

3. More specifically, in competitive industries, prices over the long term should follow the production costs, including capital costs of the marginal producer—that is, the highest-cost producer whose output is necessary to satisfy demand. While costs affect price, the reverse is also true. When the market price

falls, producers increase their efforts to reduce their costs, as the experience of the U.S. copper industry in the early 1980s illustrates.

4. Of course, there are exceptions. Ore location and other constraints may at times require mining lower-grade and poorer-quality ore early on in the life of a mine.

5. Other factors that affect reserve quality include depth of the deposit, presence of valuable by-products or costly impurities, mineralogical characteristics, and location.

6. These figures may be somewhat misleading, since some of the decline in head grade for all mines arises from the shift away from underground mining and toward leach mining over time. The high grades associated with underground mining reflect the inherently greater costs of such operations (which require high-grade ore to be profitable), while the low grades of leach mining similarly reflect its inherently lower costs. However, since the trend in average head grade for open-pit mines alone follows closely that for all mining, the magnitude of this distortion is not great.

7. The view that environmental regulations necessarily reduce productivity and increase costs, it should be noted, is not universally held. Porter (1990) and Porter and van der Linde (1995), for example, argue that environmental regulations actually increase productivity and reduce costs by fostering innovative activity. For a critical review of this position, see Palmer, Oates, and Portney 1995.

8. This and the following subsections draw heavily from Pincock, Allen, and Holt 1996; Biswas and Davenport 1994; Arbiter and Fletcher 1994; Coombs 1995; Jeric 1995; Lake 1996; Townsend and Severs 1990; and Hopkins 1994.

9. An interesting public policy issue that lies beyond the scope of this study concerns the effectiveness of U.S. government in promoting the discovery of new mineral deposits. The U.S. copper industry, as pointed out earlier, has discovered and developed very few new mines since 1970. Moreover, in recent years, the country's share of world mineral exploration has declined. These developments may reflect a deterioration in the industry's perception of the country's geological potential as well as increasingly favorable mineral polices in many countries abroad, particularly developing countries. There is, however, widespread concern within the U.S. mineral industry that the mine permitting process and other domestic regulatory policies are also discouraging the search for new mineral deposits within the United States.

10. In light of the global shift in copper exploration away from the United States and toward Latin America and other parts of the world, the U.S. copper industry in the years ahead will need to maintain a technological lead over its foreign competitors and to introduce more nonneutral innovations, like the SX-EW process, that particularly favor domestic producers if the industry is to maintain its comparative advantage and remain an important world producer. Whether it will be able to do so remains to be seen. The failure of the industry to increase labor productivity since the late 1980s is not a promising sign. However, the history of the U.S. industry shows a strong upward trend in labor productivity over the long run, but a trend that rises in steps with extended periods between each step.

REFERENCES

Arbiter, Nathaniel, and Archie W. Fletcher. 1994. Copper Hydrometallurgy—Evolution and Milestones. *Mining Engineering* February: 118–23.

Barnett, Harold J. 1979. Scarcity and Growth Revisited. In *Scarcity and Growth Reconsidered,* edited by V. Kerry Smith. Baltimore: Johns Hopkins University Press for Resources for the Future). pp. 163–217.

Barnett, Harold J., and Chandler Morse. 1963. *Scarcity and Growth.* Baltimore: Johns Hopkins University Press for Resources for the Future.

Biswas, A.K., and W.G. Davenport. 1994. *Extractive Metallurgy of Copper,* third edition. Tarrytown, New York: Pergamon and Elsevier Science.

Carter, Russell A. 1990. Kennecott Utah Copper Modernization Pays Off. *E&MJ* January: C22–C29.

Chundu, Askim, and John E. Tilton. 1994. State Enterprise and the Decline of the Zambian Copper Industry. *Resources Policy* 20(4, December): 211–18.

Coombs, David. 1995. The SX-EW Factor. *Mine & Quarry* October: 30–36.

Goldberg, Gary J. 1991. Productivity at Bingham Canyon. Unpublished paper presented by the Mine Operations Superintendent at an RTZ Mining and Exploration Conference, Salt Lake City, Utah, September.

Haberler, Gottfried. 1977. Survey of Circumstances Affecting the Location of Production and International Trade as Analyzed in the Theoretical Literature. In *The International Allocation of Economic Activity,* edited by Bertil Ohlin and others. New York: Holmes and Meier Publishers.

Hamilton, Martha M. 1988. U.S. Copper Industry is Lesson in Survival. *The Washington Post,* January 17.

Hopkins, Wayne R. 1994. SX-EW. *Mining Magazine* May: 256–65.

Jeric, Steve. 1995. The Impact of Solvent Extraction on World Supply of Copper. Unpublished paper, Colorado School of Mines, December 12.

Krugman, Paul. 1994. *The Age of Diminished Expectations: U.S. Economic Policy in the 1990s.* Revised and updated edition. Cambridge, Massachusetts: MIT Press.

Lake, James L. 1996. A Better Way to Make Copper. *Pay Dirt* December: 4–10.

McDaniel, Kirk Haley. 1989. The New Copper Range Company: Its Regeneration and the Factors Influencing It. Unpublished MS thesis, Department of Mining Engineering, Colorado School of Mines, Golden, Colorado.

National Materials Advisory Board, National Research Council. 1990. *Competitiveness of the U.S. Minerals and Metals Industry.* Washington, D.C.: National Academy Press.

Office of Technology Assessment, U.S. Congress. 1988. *Copper: Technology and Competitiveness.* OTA-E367. Washington, D.C.: Government Printing Office.

Palmer, Karen, Wallace E. Oates, and Paul R. Portney. 1995. Tightening Environmental Standards: The Benefit-Cost or the No-Cost Paradigm? *Journal of Economic Perspectives* 9(4, Fall): 119–32.

Pincock, Allen, and Holt. 1996. *Copper Technology to Year 2000*. Information Bulletin 96-1. Lakewood, Colorado: Pincock, Allen, and Holt.

Porter, K.E., and Paul R. Thomas. 1988. Competition Among World Copper Producers. *EM&J* November: 38–44.

Porter, Michael E. 1990. *The Competitive Advantage of Nations*. New York: Free Press.

Porter, Michael E., and Claas van der Linde. 1995. Toward a New Conception of the Environment-Competitiveness Relationship. *Journal of Economic Perspectives* 9(4, Fall): 97–118.

Queneau, Paul Etienne. 1985. Innovation and the Future of the American Primary Metals Industry. *Journal of Metals* February: 59–64.

Regan, James G. 1988. Miners Share $10M at Two Copper Firms. *American Metal Market* June: 27.

Stertz, Bradley A. 1989. A Once-Poor Town Gets a Lot of Advice Now That It's Rich. *Wall Street Journal,* June 29.

Tietenberg, Tom. 1996. *Environmental and Natural Resource Economics*, fourth edition. New York: Harper Collins.

Townsend, B., and K.J. Severs. 1990. The Solvent Extraction of Copper—A Perspective. *Mining Magazine* January: 26–35.

Young, Denise. 1991. Productivity and Metal Mining: Evidence from Copper-Mining Firms. *Applied Economics* (23): 1853–59.

5

Land Use Change and Innovation in U.S. Forestry

Roger A. Sedjo

This chapter looks at innovation and technical change in logging and in forest management, primarily in North America. The early history of the forest resource in the United States is one of resource depletion as the exploitation of the accessible stands has gradually given way to exploitation of less-accessible stands. However, unlike the other resources examined in this study—coal, petroleum, and copper—forests are renewable. Thus the resource depletion associated with logging and timber utilization is, in part, offset by the natural process of forest regeneration. This process has only partially offset depletion since much of the early logged forest land was converted into other uses, for instance, cropping or pasture. In recent decades, however, the process of natural regeneration has been augmented by investments, artificial regeneration in the form of tree planting, and forest plantations. Furthermore, the forest stock has been augmented by various silvicultural practices aimed at increasing forest productivity. Superimposed over the analysis are the effects of the increasingly stringent regulation of forest practices experienced in recent years, especially since the United Nations Conference on Environment and Development, held in Rio de Janeiro in 1992.

This chapter examines two sources of increased forest sector productivity. The first is technical change associated with resource extraction or logging. As forests became less available and more inaccessible, logging techniques needed to become more efficient to allow extraction at the

ROGER A. SEDJO is senior fellow and director of the Forest Economics and Policy Program at Resources for the Future.

intensive margin to continue. The second is the intensification of forest management and the increased output associated with moving to a cropping-type production mode for industrial wood.

STUDY FOCUS AND CONCLUSIONS

Although many technological innovations in forestry have taken the form of the development of new products using wood and wood fiber and new technologies for processing wood and wood fiber, the focus of this study is directed at the raw resource itself and on important innovations in harvesting and growing industrial wood fiber. Figure 5-1 describes the flows from the raw wood resource. Some of the wood succumbs to various forms of natural mortality, such as fire and infestation. Humans harvest wood for use both for fuel and for industrial wood, that is, wood that is processed into wood materials and woodpulp for paper. This study is concerned with the approximately 50% of raw wood that is harvested worldwide for use as industrial wood.

The study focuses on (a) technological innovation in wood extraction (logging) and (b) technological and institutional innovations in tree growing, with the focus being on intensive management of plantation forests. The examination takes place within the context of changing societal attitudes toward forests as reflected in changing regulations and laws regarding logging and forest practices.

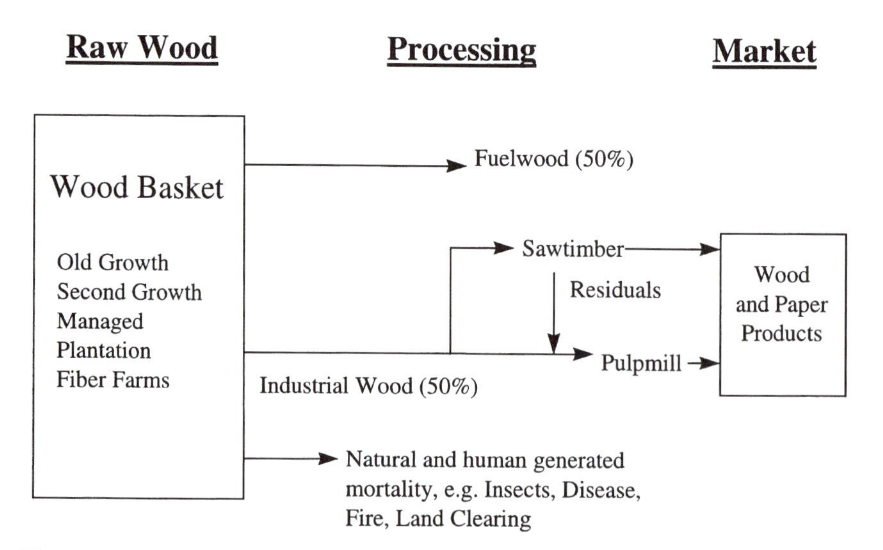

Figure 5-1. Wood Flows and Uses.

The study examines these innovations in the context of recent land use policies that have affected and are affecting industrial forestry and the production of industrial wood. The perspective is primarily that of North America, the dominant wood-producing region of the world, although not all the examples are from North America. The forest management focus is on prospects in the United States, both with traditional intensive forest management and the even more intensive fiber farming.

The study concludes that the single dominant innovation within the forest industry in recent decades is the shift to intensively managed forest plantations and fiber farms. The advantages of plantations relate to control of species and location, to the achievement of higher productivity through management,[1] and to the opportunity to capture returns to genetic improvement through planting of superior seedlings that increase tree growth and allow for control of the quality of tree and the tree fiber. Furthermore, intensively managed forest plantations have been demonstrated to be especially favorable on high biological sites, which can support rapid tree growth, and they can be established in locations with access to processing facilities and to markets, both at home and abroad.

The basic trend toward an expansion of investments in intensively managed plantations is being accelerated by changing land use policies, which preclude harvests on some lands and raise costs on others. Obtaining wood from the extensive margin (that is, from old-growth forest) is becoming increasingly difficult, costly, and uncertain as old-growth forests become inaccessible due to decreasing physical availability and also due to the increasing pressures for protection being applied by the environmental community. The ability of intensive forest management to increase yields five to ten times the natural forest growth rates has provided firms with the opportunity to conserve their expenditures on land ownership and focus their forestland holdings into a smaller, more productive area. The recent development of the fiber farm reflects what is perhaps the most advanced form of intensive tree management, rivaling that of much of agricultural cropping.

In harvesting, this study found that there has been a stream of new equipment and techniques that are being applied in most of the regions examined. However, there has been no single large innovation that dominates the forest industry. Rather, a number of small innovations have resulted in gradual changes in harvesting that are maintaining a stable labor productivity. The stability of aggregate productivity may hide the large variability in technologies used in harvesting, even within a region. This study also concludes that harvesting labor productivity has increased slightly in recent decades, suggesting that extraction and regrowth technology has kept slightly ahead of harvestable resource depletion, some of which has been caused by environmental constraints and forest set-asides.

BACKGROUND

Industrial Wood Uses

Industrial wood has two major uses: for building materials, such as those used in construction, furniture, and the like, and for fiber products, typically various types of paper and paperboard used for writing, printing, and packaging (Figure 5-1). Although wood for construction and materials has been used for millennia, the use of wood fiber for paper production has only been common for about the last century, since previously paper was made out of plant and animal fibers and often from old cloth. Traditionally, hardwood and conifer have been used to produce quite different types of products. Initially, paper was quite limited in its use of species due to concerns about both wood fiber characteristics and wood resins. For example, due to the nature of pine resin, it was not until the post-World War II era that Southern pine was commonly used to produce newsprint. Additionally, the length of the fiber is important to the strength in paper, and conifer was preferred for its long fiber. Thus, short-fiber hardwood had only limited use in paper production.

While for some construction purposes various types of wood were interchanged, there were still preferences based on milling and nailing characteristics, surface characteristics, and visual features. Also, with the advent of plywood and other panel materials, certain types of woods had advantages for the production of certain products. Also, gradually, the short fiber of various deciduous species was mixed into the process both to impart paper with a smoother surface, more amenable to writing and printing, and also as a less-expensive filler material. Today, many pulpmills face scarce supplies of certain short fibers due in part to new packaging approaches that require more printing on the package surface and also because much of the hardwood resource that exists is in less accessible, often wetter, areas.

In recent years, however, fiber products have come to include not only paper products, but also a host of solid materials that have been produced by the composition of wood fibers. These include various types of reconstituted fiberboard and engineered wood products. Commonly, tree species is important and historically certain species have physical characteristics that allow for their use in producing specific products. However, through time, technology has allowed increasing degrees of flexibility in utilizing substitute species.[2]

Forest Exploitation and Extraction

Until recent decades, humans could move on to new forested areas as the areas exploited earlier become unavailable. However, as natural forests become increasingly inaccessible, financial decisions at the margin are shifting increasingly to investments in intensive management and plantation forests, which can generate positive financial returns. Today, one can view the globe as being in a transition in the provision of its industrial wood resource needs away from a gathering-and-foraging mode in natural forests to an agricultural mode of husbandry and cropping.

In the United States, forest resources were drawn first from New England, the Mid-Atlantic states, and parts of the Southeastern Coastal Plain; then from the Ohio Valley, the Lake states, and the South Central states; and finally from the Pacific Northwest. Canals and later railroads added to the natural transport network provided by rivers. The U.S. forest frontier was gradually extended into Canada, and Canadian wood became an increasingly large component of U.S. consumption.[3]

Worldwide, new sources of wood have been developed as traditional sources became depleted or the forestland was converted to other uses. The natural tropical forests of the Philippines, Malaysia, and then Indonesia became the leading suppliers of tropical hardwoods for world markets only in the post-WWII period, and for Indonesia only since the 1970s. Thus, the forests of Japan were complemented by wood from the Asia-Pacific region, as well as from North America and Russia. The vast forest of Russia was an important source of industrial wood for the centrally planned economies as well as portions of the Pacific Basin and European markets until the Soviet Union's political breakup in 1989 and the early 1990s led to disruptions that have adversely affected timber harvests. Similarly, harvests from Canada's natural forest have expanded as world markets grew, and Canada has become the world's largest industrial wood exporter.

In many respects, however, the exploitation of the last of the major frontier forests of the world, with their huge inventories of natural mature forest, may already have been undertaken. Globally, there are few new native forests that can be developed or exploited. The largest remaining forest areas that have not yet been developed for large-scale commercial logging are found in the Amazon and the Russian forests of Siberia.

The Current Situation: Industrial Wood Sources

The industrial wood resource comes from a host of regions around the globe. Table 5-1 provides a global overview of industrial wood production in a recent year. The data reveal that although North America is an important supplier of industrial wood, accounting for more than 37% of global production, there are also many other important suppliers.

Today, a variety of sources form the world's wood basket. Native forests are the primary source of industrial wood supplies in almost all of

Table 5-1. World Industrial Wood Production, 1993.

Country	*Volume (cubic meters)*	*Percent Total*
United States	402,500	26.3
Canada	173,133	11.3
Europe, excluding Nordic	169,036	11.1
Nordic	103,213	6.7
Russian Federation	172,955	11.3
Japan	32,209	2.1
Indonesia/Malaysia	84,011	5.4
China/India	125,275	8.2
Other	—	17.6
Total World	1,529,000	100.0

Source: UN FAO Forest Yearbook, 1995.

Canada. Much of the Russian harvest is drawn from virgin forests, especially in Siberia and the Russian Far East. Although vast forests remain in Siberia, they are unlikely to provide a major source of industrial wood in the foreseeable future due to their economic inaccessibility. Under the old Soviet system, these forests could be exploited, since the huge transport costs to market were subsidized. However, it is problematical whether significant volumes of Siberian industrial wood can be transported economically to major world markets if the price must fully bear the transport costs, although the forests of the Russian Far East are much more accessible to international transport.

Industrial wood from tropical forests accounts for less than 25% of the world's industrial wood production and comes largely from logged areas of older forests, with southeast Asia and the Asia-Pacific region being a major tropical wood source. Due to the limited number of commercial species found in most tropical forests, commercial logging in tropical forests almost always takes the form of selective logging whereby only a relatively modest portion of the trees are felled and removed.[4] The Amazon holds huge remaining wood stocks; however, these forests are difficult to utilize as a major source of industrial wood given current technology since the species are so diverse and varied that large-scale commercial logging of the native forest has not been economically feasible.[5] Furthermore, pressures by environmentalists will make commercial harvests in the Amazon more difficult, even should technology adapt. The deforestation occurring in the Amazon is largely driven by forest conversion to agricultural uses, and little of the wood is utilized as industrial wood.

In much of Europe and the United States, the timber harvests are drawn from second-growth forests. In Europe, the forests have often been replanted to ensure regeneration, while in much of the United States, the harvests are drawn from areas that were logged over and have regenerated naturally or from areas that were formerly agricultural lands that had been neglected and naturally regenerated.

In addition, a growing portion of the world's industrial wood is being supplied from plantation forests. Although many of these are in traditional forest areas (including parts of Europe and the U.S. South), a growing portion of the world's industrial wood basket is coming from plantations in nontraditional wood-producing regions. These regions include several countries in South America (Brazil, Chile, Venezuela, Uruguay, and Argentina) as well as countries such as New Zealand, Australia, South Africa, Spain, Portugal, Indonesia, Thailand, and China. All of these countries have important industrial tree-growing activities, and several have become major wood producers, not only for their domestic markets, but for international markets as well.

Finally, a new source of wood fiber is being developed in the form of the *fiber farm*. The fiber farm takes plantation forestry to its logical extension in that fiber is grown as an agricultural crop in very intensively managed short rotations on sites that typically until recently were used for agricultural crops. Where water is lacking, it is provided through drip irrigation, and where nutrients are lacking, they are provided through appropriate fertilization. Fiber farms are a subset of intensively managed plantations and differ only in the degree of the intensity of management. Although this form of plantation is still in its fledgling stages of development, probably occupying less than 40,000 hectares currently in the United States, this type of activity has increased substantially in the past decade.

A crude disaggregation of the types of forests from which current timber harvests are supplied is provided in Table 5-2. This table suggests that approximately 30% of current timber harvests come from old-growth forests, 10% from exotic plantations, and 60% from second-growth forests, most with some degree of management.

LOGGING

Timber extraction or logging practices vary depending upon the species, tree size, location, and terrain. Figure 5-2 provides a schematic of alternative logging situations and provides an estimate of the relative logging costs likely to be experienced for the various forest situations. As the figure shows, in general, it would be expected that logging costs would be highest in old-growth forests in difficult terrain and lowest in intensively managed plantations. Not only is terrain important, but stand homogene-

Table 5-2. Global Harvests by Forest Management Condition.

Forest Situation	*Percentage of Global Industrial Wood Harvest*
Old-growth	30
Second-growth:	
Minimal management	14
Indigenous, managed	22
Industrial plantations, indigenous	24
Industrial plantations, exotic	10

Notes: Old-growth includes Canada, Russia, Indonesia/Malaysia. Second-growth, minimal management, includes parts of the United States and Canada, Russia. Second-growth, indigenous, managed, includes residual. Second-growth, industrial plantations, indigenous, includes Nordic, most of Europe, a large but minor portion of the United States, Japan, and some from China and India. Industrial exotic plantations, see Sedjo 1998.

Source: Personal estimate drawing from one-country production levels provided by the FAO.

ity and volumes of timber per unit of land area also are significant influences. Also, selective logging—for instance, the logging of only a few of the more desired trees—can increase harvesting costs. Thus, forest plantations and fiber farms would be expected to have the lowest harvesting costs. An advantage of forest management is that it can influence growth so as to generate high volumes of homogeneous trees on accessible sites on easy terrain.

Elsewhere (Sedjo 1997) we examine in greater detail modern logging technology in four regions with very different conditions, running the gamut from flat terrain and small trees, to steep terrain and large trees. The Nordic forests examined are characterized by relatively flat terrain, small trees, and long periods of frozen soil, which allows for ease of logging. The area of the Pacific Northwest examined was one of naturally regenerated second-growth forest, largely hemlock, moderate-size trees, relatively flat terrain and high levels of precipitation. Precipitation creates soils that are wet for large portions of the year, making logging more difficult, in part due to greater soil damage associated with logging on wet soils. The coastal forests of British Columbia are characterized by steep terrain, wet weather, and relatively large old-growth trees. Finally, the U.S. South is characterized as having large areas of flat terrain, moderate size trees, and substantial periods of the year with wet soils.

Variations in Logging Innovations across Countries[6]

Innovations in logging appear to draw largely from technologies[7] developed in other similar industries, which are then adapted and modified to

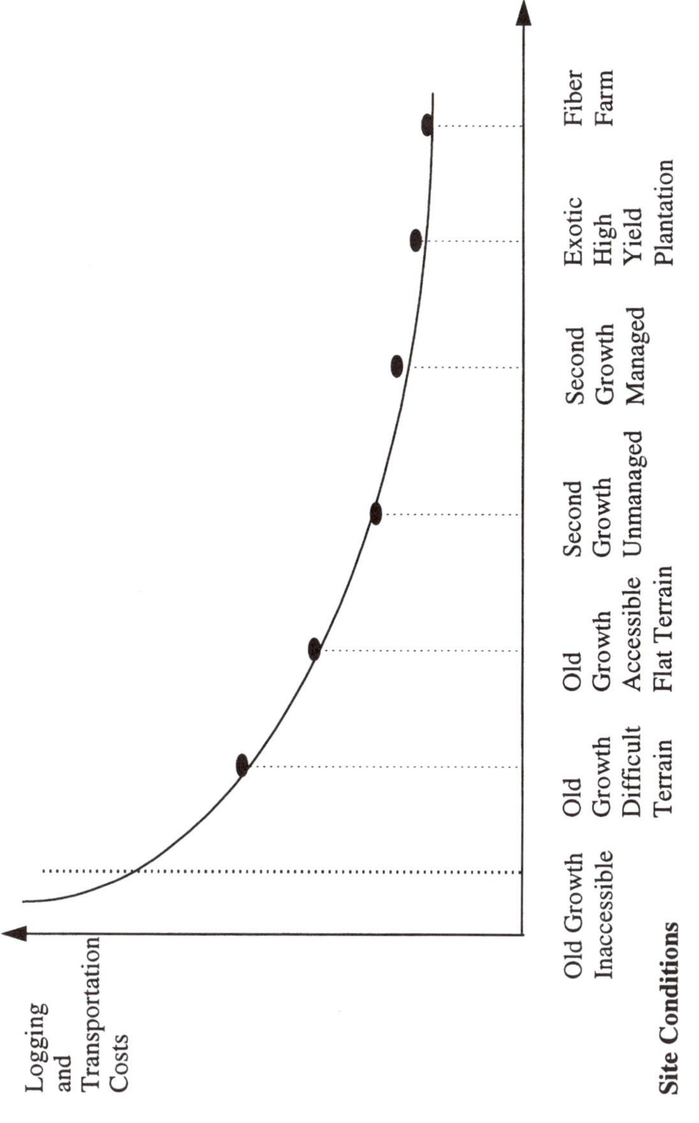

Figure 5-2. Hypothetical Logging Costs: Alternative Site Conditions.

meet the specific needs of logging. Perhaps the most important post-WWII innovation in logging was the chainsaw, which substituted mechanical power for human muscle in driving the saw to fell and cut the tree. This was accompanied by the development of increasingly sophisticated power equipment (which replaced draft animals) for removing the logs from the forest.

Both of these innovations drew upon technologies developed elsewhere. The small powerful internal combustion engine was adapted for the chainsaw, and tractors and transport equipment were drawn from agriculture and heavy equipment manufacturers. In the last several decades, there does not appear to be any single dominant technological innovation in logging that has had the impact of the chainsaw. Rather, in many forests, the chainsaw has been replaced by mechanical tree fellers and other power equipment that have subsequently been replaced by the second, third, and fourth generations of improved power equipment.

The logging industry has evolved different harvesting styles in Canada, the United States, and Sweden. Although some of these differences, no doubt, reflect different site, climate, and locational conditions for logging, the path also reflects different histories and social forces. In general, the changes in logging reflect not only site conditions and tree sizes, but also the relationship between the processing functions that are done in the mill vis-à-vis those done in the forest. In the Nordic countries, for example, the earlier post-WWII period was one when attempts to maintain populations in the rural northern area were breaking down. Labor was no longer willing to endure the hardship of the logging camp. The decline of the small farm further reduced the labor force available for harvesting, even for seasonal employment in the winter months. In this context, government policy directed the development of mechanization that performed the same tasks that had been done in labor-intensive systems. Expensive labor-saving equipment, developed by the government, included innovations tending to mimic earlier human activities such as felling, bucking, forwarding the wood to the pick-up point, and so on (Laestadius 1990). Equipment with booms, which could reach out for a tree, was required since glacial detritus often precluded driving to the tree, and the orientation was to deliver the wood to the primitive roads that existed. Forwarders were developed to mimic crude equipment used by forester-farmers.

Canada, too, had difficulty recruiting labor in areas far from urban centers. However, the major transport to mills was via water routes, often at a substantial distance from the logging. A system of manual felling and tree-length skidding to rivers was used where water transport was available. Gradually, trucks replaced water transport, except on the Pacific coast.

By contrast to the Nordic countries and Canada, the post-WWII U.S. South had large numbers of rural communities and was able to utilize the agricultural labor pool during nonpeak periods in agriculture. Logging provided attractive off-season employment to the essentially agrarian labor force. Throughout much of the postwar period, this seasonal labor was employed, often self-employed, in logging and transport. Technical requirements for logging were minimal, and the farmer-entrepreneur would also provide a tractor, truck, chainsaw, and so forth. Thus there was no need for wood-consuming firms to invest in harvesting, as numerous small contractors would provide the service of buying, harvesting, and delivering the wood to a pickup point, often a rail siding, or directly to the mill.

These different labor conditions have led to the development of local institutions that have impacts on harvesting operations and variability. The Swedish harvester has little excess capacity since its harvests are predetermined by annual targets.[8] Its specialized equipment provides a continuous flow of wood to the mill, where it accumulates as inventories (Laestadius 1990). In the United States, by contrast, changes in the mill's wood requirements are met to a much greater degree by changes in local logging activity. The Southern contract logging system, which evolved in an environment of seasonal labor and equipment drawn from agriculture, has an "excess capacity" feature that allows logging production to vary so that wood inventories are small and wood flows are "just-in-time."

As the post-WWII period continued, however, logging in North America and especially the South has been modified following the demise of the small family farm and the large-scale migration to urban areas. In recent periods, many, if not most, of the innovations have taken the form of labor-saving technology (Stier 1982; Cubbage and Carter 1994) as labor has become the scarce factor and real wages have been increasing. Nevertheless, the contract logger still continues to dominate, and opportunities exist for small logging operations staffed by seasonal employees.

Aggregate Productivity Growth in Logging

Anecdotal evidence from all of the regions examined suggests that in recent periods labor has generally been the scarce factor and other inputs, such as capital and labor-saving technology, were being substituted for labor. This interpretation is consistent with rising real wages that have been occurring in each of the regions in the post-WWII period.

In a survey piece, Stier and Bengston (1992) reviewed technical change in the North American forestry sector. Although they mentioned a large number of studies and discussed in detail twenty-four studies that examined total factor productivity, most of these studies examined some part of the wood-processing activity, such as lumber milling or pulp and paper making. Only one of these (Stier 1982) focused on logging and wood extraction. Stier's 1982 study is consistent with the anecdotal evidence. He found that wages in North America increased threefold and the capital-labor ratio more than doubled over the 1950–1974 period examined, indicating the large substitution of capital for labor. Furthermore, he concluded that the "substitution away from labor has taken place primarily through changes in technology in the form of new labor-saving equipment" (265). A separate study related to wood extraction by Cubbage and Carter (1994) found that the costs over time associated with the more labor-intensive shortwood pulpwood harvesting systems in the South generally increased, while those of the labor-saving longwood harvesting systems declined.

Parry (1997) examined total factor productivity growth for the United States in the natural resources industries between 1970 and 1994 using aggregative data. For U.S. forestry, Parry found that labor productivity rose about 20% from 1970 to 1993 (Figure 5-3). This compares with the estimated 33% increase in labor productivity in coastal British Columbia for the forest industry (Sedjo 1997) and with estimated labor productivity growth in U.S. manufacturing over that same period of about 34% (Parry 1997).

An examination of multifactor productivity, however, gives a different picture. During the period 1970–1993, Parry found that multifactor productivity in U.S. forestry declined about 20% (Figure 5-4). These results suggest that labor productivity grew due to the introduction of nonlabor inputs into forestry production and their substitution for the labor inputs. However, the results also indicate that the overall increase in productivity from the introduction of improved technology has not been adequate,[9] overall, to offset the decline in the accessibility[10] and the quality of the resource.[11]

PLANTATIONS AND INTENSIVE MANAGEMENT

Millennia ago, humans made a transition from gathering and hunting to cropping and herding. A similar transition is occurring in forestry today as "cropping" tree-growing plantations are expanding as "gathering" old and second-growth harvesting declines. Although much of the transition is driven by economic and social considerations, the advent of technology that can be used in plantation forestry facilitates the transition. While many of the basic cropping techniques are simply the application of agri-

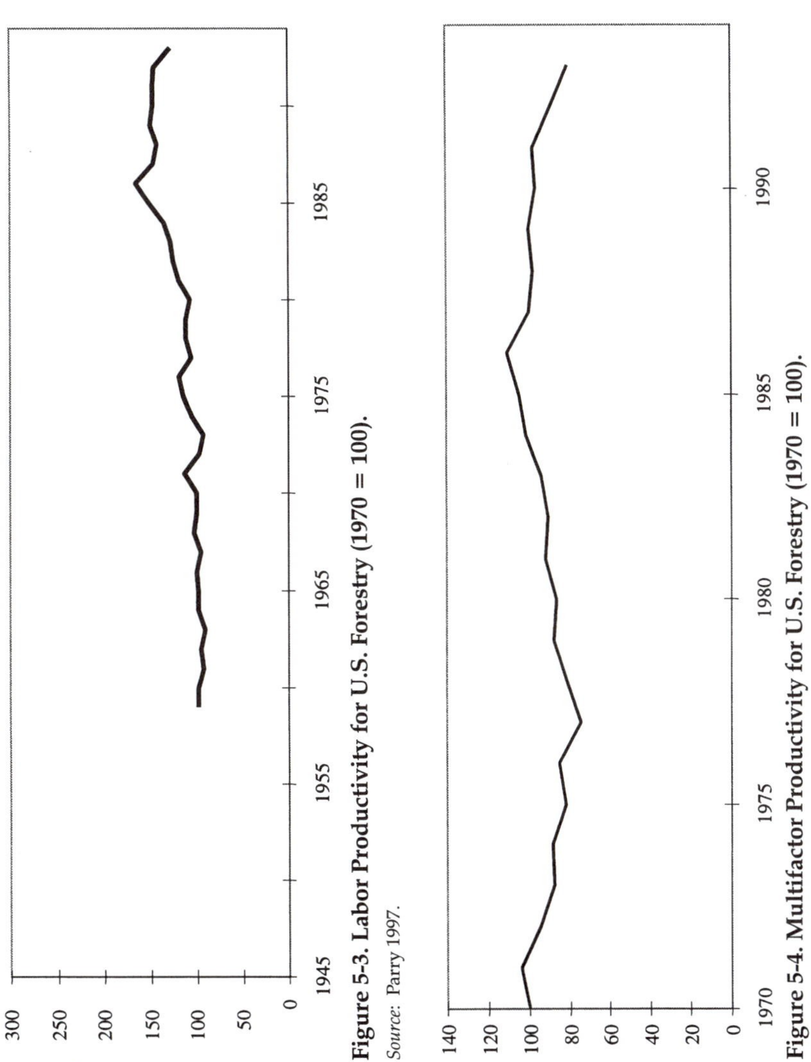

Figure 5-3. Labor Productivity for U.S. Forestry (1970 = 100).

Source: Parry 1997.

Figure 5-4. Multifactor Productivity for U.S. Forestry (1970 = 100).

Source: Parry 1997.

cultural approaches to trees, not all the technology transfers are accomplished easily or unmodified. Wild trees need to be domesticated, as did wild maize or wheat. Furthermore, a stock of knowledge must be accumulated on the characteristics of trees in general and the various species. Additionally, improvement in the genetic stock of trees involves the modification and adaptation of genetic techniques developed outside forestry.

The Trend toward Tree Plantations

At the margin, total timber production can be expanded by (a) increasing harvests from native forests; (b) increasing forest management in second-growth and other forests; (c) expanding the area of tree planting and plantation forest; or (d) increased intensity of forest management on plantation forest.

In most parts of the world, trees will regenerate naturally on forest-land unless prohibited from doing so by land-use changes, such as to cropping or pasture, or by excessive and continuous land disturbances. Although the history of planting trees and forest management goes back millennia, it is only within the past few decades that industrial tree planting has become an important economic activity. Much of the tree planting that occurred earlier in North America represented environmental concerns and the creation of protected forests for watershed purposes, to prevent avalanches, or as part of land rehabilitation schemes as with the Civilian Conservation Corps during the Depression years. Elsewhere, most tree planting was also associated with environmental concerns and land reclamation, as in the Netherlands for the creation of new lands from the sea.

Beginning during the 1960s, however, major industrial tree plantation activities began to be undertaken in a number of regions of the world, including the Pacific Northwest and the U.S. South, as well is in other regions such as New Zealand, Chile, Brazil, and South Africa. In the U.S. South and the Pacific Northwest, where natural regeneration often resulted in the growth of less commercially valuable hardwood species, tree planting also allowed for regeneration of the desired species, as well as the assurance of rapid and complete forest regeneration.

Advantages of Plantations

The advantages of industrial forest plantations are similar to the advantages of agriculture cropping over foraging. For example, tree planting and intensive management allow for the control of inputs to achieve higher growth rates even without genetic improvements.[12] Additionally, control over the location of the plantation allows for the choice of lands where (a) the terrain is favorable to planting and harvesting activities, (b) soil and climate are conducive to higher biological productivity, (c) species can be chosen to fit the site conditions, and (d) the area is accessible to mills and markets. As in agricultural cropping, uniformity in product as well as other favorable conditions associated with the site allow harvesting costs to be lower than in most natural forests.

Tree planting per se has the advantage of allowing for the choice of species and therefore the choice of type of wood fiber (short or long fiber) as well as the possibility of the introduction of fast-growing species that may not occur naturally on a site. For example, various types of pine have been introduced into South America, New Zealand, and Australia—areas that do not have indigenous pine. In many cases, the growth of pine in its new environment has exceeded that in its natural environment,[13] and pine plantations are now common in many parts of these regions. Similarly, eucalyptus, an Australian genus, has been adopted and flourishes in many regions while exhibiting excellent growth characteristics. Also, planting allows for intensive management—such as fertilization, pest control, and thinning—to be undertaken throughout the growth cycle to promote high yields and desired characteristics.[14] As in agriculture, evidence suggests that high-intensity management generates the most favorable financial returns on the high-productivity sites and relatively lower returns on less favorable sites (for examples, see Sedjo and Lyon 1990). Finally, as with agriculture, planting allows for the introduction of genetic improvement through the selection of seed and seedlings from superior trees, hybridization, and so forth. Without planting, investments in genetic tree improvement have no commercial applications.

A recent study (Sedjo 1998) estimated that the share of the industrial wood production from plantations in the tropics and subtropics, using almost entirely exotic (nonindigenous) species, has roughly doubled in the fifteen-year period from 1977 to 1992, from about 5 to 10% of the world's total production.[15]

GROWTH, BIOTECHNOLOGY, AND GENETICS[16]

Some Research Findings

In an assessment of the effects of research on the productivity of the forestry sector, Hyde, Newman, and Seldon (1992) corroborated the work of earlier researchers reviewed by Stier and Bengston (1992) when they found substantial economic returns to a number of research activities related to the development of new products and new technologies in wood processing. The Hyde and others study also assessed the benefits of research specifically in wood growing, such as timber growth and forest management, on softwood timber production in the U.S. South. Earlier, Newman (1987) had estimated that aggregate productivity in a southern composite of softwood forest inventory plus removal increased at an average of 0.5% to 1.0% annually over the period 1935–1980. However, Hyde and others found little positive economic return, even though aggregate productivity of the forest was increasing, and they concluded that "Net present value and the internal rate of return estimates are uniformly poor...." They continue, "Apparently, research benefits in southern softwood growth and management have not led to large social gains" (192).

One reason given by Hyde and others (218) for the lack of apparent net economic benefits to forest management and tree planting was the large overhang of old-growth forests elsewhere in North America and the reality that technological improvement in Southern pine is probably justified only on a subset of industrial lands. Alternatively stated, it makes little sense to invest in activities that produce more rapidly growing trees when there are large volumes of mature timber that are available for the cost of the logging. This explanation is similar to that of Hayami and Ruttan (1971, 115) for U.S. agriculture, where they found that investments in yield-improving technologies were not forthcoming until essentially all of the potentially usable agricultural land was in agricultural use. Until all of the land available for cropping was in use, investments in equipment, designed to extend the area of land that could be cropped, generated the highest returns. However, once the frontier had disappeared and new crop land was not readily available, returns to research investments that increased agricultural yields began to be substantial.

There is now some anecdotal evidence, found in the behavior of forest firms in recent periods, that this threshold may have been passed. Many forest product firms are investing substantial sums in tree-growing research efforts, and biotechnology research is common in many intermediate to large firms. Additionally, many industrial forest products firms appear to have a long-term policy of consolidation of the best lands for tree growing combined with an active genetic improvement research program. For example, Westvaco Corporation has as its explicit long-term strategy the objective of consolidating its land holdings to those of highest productivity and then managing intensively, including genetic tree improvement. Thus, a major segment of the market is behaving as if it believes that the long-term returns to research in areas designed to increase returns to intensive management are substantial.

This interpretation suggests that as the old-growth forest becomes less available (due to earlier exploitation, access problems, or political considerations) and experiences higher costs, the high-return investments would be expected to be found in research that generates yield-increasing technologies. It also suggests that, historically, higher returns were likely to be associated with innovations that reduced the costs of logging and transport to the mill. Thus, perhaps the findings of Hyde and others that research in tree improvement found little support are not surprising since, until recently, those improvements could not be introduced into wood fiber production in a major way.

Firm behavior is now suggesting that the United States, indeed the global forest industry, is moving through a transition from primary reliance for meeting industrial wood needs on naturally generated forests to increased reliance on planted forests. One might view plantation forests as providing the technological backstop that constrains the price growth of industrial wood as wood "scarcity" can be offset by investments in plantation forests (see Solow 1974). To the extent that the real price of industrial wood continues to rise, the returns to investments in forest plantations increase.

Further accelerating this trend are the enhanced returns to recent genetic innovation that are occurring as the wood fiber from natural forests is becoming even more inaccessible as the remaining unlogged forests are increasingly set aside to increase the area of protected forests in various protected reserves. This change should increase the attractiveness of several yield-increasing activities, including forest management, industrial forest plantations, and genetic improvements that increase tree growth and improved tree and fiber characteristics.

Tree Improvement Programs

With the planting of trees come the incentives for tree improvements. Tree improvements can take many forms, including:

- growth rates,
- disease and pest resistance,
- tree form and wood fiber quality, and
- desired fiber characteristics that relate to ease in processing.

The most common emphases of tree improvement programs are increased growth, disease resistance, and wood quality. *Growth* typically refers to wood volume growth or yields. *Disease and pest resistance* is usually oriented to specific problems common in a particular species. *Wood quality* includes a variety of characteristics including tree form and wood fiber quality, such as straightness of the trunk, the absence of large or excessive branching, the amount of taper in the trunk, and so forth. These features are related to the usefulness of the wood in the production of products in the final output. *Desired fiber characteristics* may relate to ease in processing, including the breakdown of wood fibers in chemical processing; reduced lignin, which provides more desired pulping properties; and the extent and ease with which the wood can be converted into woodpulp and paper products with desired characteristics or properties, such as paper tear strength, surface texture, whiteness, and so forth. In addition, wood fiber is increasingly being processed into structural products such as strand board, fiber board, and engineered wood products, which have their own set of desired fiber characteristics.

In recent years, pulp producers have begun to move away from simply producing a standardized "commodity" pulp and are producing specialized pulp for targeted markets. Aracruz, a Brazilian pulp company, has asserted that it can customize its tree fibers to the requirements of individual customers. This requires increased control over the mix and types of wood fibers used. Customized products often require customized raw materials.

Biotechnology and Genetics

Biotechnology comprises any technique that uses living organisms to make or modify a product, to improve plants or animals, or to develop microorganisms for specific use (Haines 1994a, 1994b). In the tree-growing industry, genetic research takes two forms. First is the use of traditional breeding techniques that utilize the variation in the natural population to breed trees with desired traits. These techniques also include hybridization techniques, which have provided more robust offspring by bringing together populations that do not normally mix in nature. These techniques have proved very successful in agriculture, and their application to forestry is already well developed. One difficulty with traditional approaches in tree breeding, however, is the long growth cycle that makes this process very slow. However, techniques that can help identify trees with desired genes, such as molecular biology and genetic marking, can enhance the speed of developing genetic improvements.

The second form of biotechnological research uses genetic engineering approaches whereby genes with desired traits are artificially introduced into a species that could never acquire these genes through natural processes. This approach is still in its early stages but is more well developed in agriculture than forestry. Nevertheless, applications to forestry are occurring, and the potential appears great.

The success of genetic approaches for increasing yields and generating desired modifications in plants is unquestioned in agriculture. Traditional breeding techniques have resulted in the development of a large array of improved seeds that have generated yield increases in a host of agriculture products. Furthermore, applications of genetic engineering in agriculture have been successful in transferring to agricultural crops desired traits such as resistance to disease and herbicides, resistance to cold, and so forth. The same possibilities are available for trees, and the existing techniques are being modified for application to trees.

Genetics: Some Applications to Trees

As in agriculture, hybrids are now being developed in forestry. Hybridization crosses trees that are unlikely to breed in nature, often where parents do not occur together in sympatric populations. These crosses often exhibit growth and other characteristics that neither of the parent species alone can achieve. In the United States, for example, hybrid poplars have shown remarkable growth rates exceeding those found in the parent populations,[17] and pitch pine–loblolly pine hybrids exhibit the cold hardiness found in pitch pine and the rapid growth of loblolly pine.

In recent years, molecular approaches to tree selection and breeding have shown great promise. The molecular approach typically involves genetic material being identified, collected, bred, and tested over a wide range of sites. Rather then simply choosing specific tree phenotypes on the basis of their characteristics, the molecular approach identifies the chromosomes that are associated with the desired traits. Markers are then used to identify these genes in the collection/screening process. This approach allows for exploitation of the abundant genetic variation found in natural populations. When a particular gene or set of genes has been found to be associated with a desired trait, the use of molecular markers and screening techniques allows for the examination of the DNA of thousands of individual trees to identify the few, perhaps less than a dozen, with the optimal mix of genes. These techniques are currently being applied to the development of improved poplars in the United States and eucalyptus in Brazil (Bradshaw 1997; Westvaco 1997).

Recent work using molecular genetics on hybrid poplar in the Pacific Northwest has shown a 20% increase in yields in plantations and an additional 20% on dry sites where irrigation can be applied (east of the Cascade Mountains) (Bradshaw 1997). Also, improvements in yields continue (Withrow-Robinson, Hibbs, and Beuter 1995, 13). Growth rates with these plantations are impressive, with yields of about seven tons per acre, or about fifty cubic meters per hectare. These growth rates are three times the growth rates on typical pine plantations in the U.S. South. In some cases, as in Aracruz in Brazil, yields of hybrid eucalyptus are reported to have more than doubled.

Finally, research is underway to examine the potential of using genetic engineering for producing superior trees. Genetic engineering allows novel genes to be transferred among species that cannot be transferred in nature and the modification of existing genes through manipulation of the DNA molecule. Currently, there are discussions underway to introduce a gene into plantation trees that creates a resistance to the herbicide glyphosate (trademarked as Vison, Accord, or Roundup). This approach, now practiced in some agriculture, would allow for the planted area to be sprayed with the herbicide without damage to the tree, thus ensuring a fast and successful start to tree growth.[18]

As the ability to introduce meaningful tree improvements increases, the financial returns to plantations are likely to improve vis-à-vis those of natural forest harvesting, especially as the old-growth overhang is reduced. In general, traditional breeding approaches have been shown to produce substantial yield improvements (see Table 5-3). For example, an orchard mix of first-generation open-pollinated seed can be expected to generate an 8% per generation improvement in the desired characteristic, such as yield. Other seed collection and deployment techniques, such as choosing the best mothers (Family Block), can result in an 11% increase in yield, while mass-control pollination techniques, which control for both male and female genes, have increased yields up to 21% (personal communication with Westvaco researchers, Summerville, South Carolina, December 17, 1996).

Clonal Applications

An approach for transferring genetic traits and developing large numbers of improved seedlings is through *tissue culture*. Tissue culture can produce multiple copies of an elite genotype as well as provide a means of introducing novel genes. In one approach, called *organogenesis*, plant tissue is placed on a nutrient medium and buds are initiated on the plant tissue. From these buds plants are developed. Where elite genetic material is available, in vitro plant propagation (through various tissue and cell culture techniques, *micropropagation*) has the ability to multiply clonal material very rapidly. Such tissue culture techniques provide the tools to produce genetically engineered plants and to regenerate trees with

Table 5-3. Gains from Various Traditional Breeding Approaches.

Technique	*Effect (% increase in yields)*
Orchard mix, open pollination, first generation	8
Family block, best mothers	11
Mass pollination (control for both male and female)	21

Source: Westvaco Corporation.

desired traits quickly, without having to wait for the trees to reach sexual maturity (Westvaco 1996, 8–9).

In other cases, such as the use of seed orchards and conventional breeding techniques, cloning approaches are limited and more traditional approaches must be used. Using modern cloning techniques, thousands of clones with the desired traits can be produced for later establishment in tree or fiber farms. However, for various technical reasons, cloning is more difficult with conifers than with deciduous trees.

The development of clonal approaches to propagation is important to the broad utilization and dissemination of genetically improved stock. With tree planting often involving more than 500 seedlings per acre,[19] large-scale planting of improved stock requires some method of generating literally many millions of seedlings, at relatively low cost, that embody the genetic upgrading. The costs of the improved seedlings are important in a financial sense since the benefits of the improved genetics are delayed until the harvest. With harvests often being twenty years or more after planting, large costs for improved seed are sometimes difficult to justify financially. However, if the costs of plantings are going to be incurred, the incremental costs associated with planting improved genetic stock are likely to modest and thus more likely to be financially justified.

The Effect of Genetics on Fiber Production

Thus far, tree improvement programs have probably had relatively little effect on fiber production in North America since there has generally been at least a twenty-year lag between the introduction of an improved seedling and the capture of that increased productivity in a harvested tree. In the United States, most genetically improved trees that were planted during the 1980s will not be harvested until after 2000. However, the advent of fiber farms brings the possibility of a much faster impact on production, since rotations are as short as six years. Similarly, industrial forest plantations in the semitropics, using eucalyptus, often have rotation periods as short as six or seven years. With short rotations, the impacts of new genetic technology embodied in seedlings will have a much more rapid effect on industrial wood volumes.

An anecdote indicates that tree growers believe that technological improvement in seed stock is important. In some places where eucalyptus trees had been planted, they have been replanted after harvest. The original expectation had been that the single planting would generate three harvests through the natural process of *coppicing*, a characteristic exhibited by many deciduous tree species whereby the stump of a felled tree will react by sprouting new limbs, which can grow into trees. Initially, the plan was to rely on coppicing in order to save the costs of replanting. However, in fact, this has rarely been done. Rather, the stumps are treated and new, genetically superior stock is planted. The rationale for the replanting is often because the genetic improvement is viewed as justifying the planting of the superior stock, rather than relying on the genetics of the existing tree.

Fiber Farms

In the past decade or so, as noted above, fiber farms have been introduced in the United States and Canada. Fiber farms differ from plantation forests in the degree of management intensity. In the Pacific Northwest, they are often established on prime agricultural lands or on irrigated arid or semiarid lands east of the Cascades, but typically not on recently logged land or on marginal agricultural lands that have simply ceased agricultural production. In the Pacific Northwest and British Columbia, forest products firms have established 20,000 to 25,000 hectares, largely in hybrid poplar (which is a cross of black and eastern poplar), with rotations of six to twelve years. One advantage of the short rotations is that if they are ten years or less, state law in Washington and Oregon treats the operation as agriculture rather than forestry, and the regulations applying to agriculture are typically less stringent than those applying to forestry. In British Columbia, the legal requirement for agricultural treatment of a fiber farm is a rotation of twelve years or less. Poplar plantations are common in Europe; however, they tend to occur in small plantings and with longer (fifteen- to twenty-five-year) rotations.

In areas of high precipitation, as west of the Cascades in the Pacific Northwest, hybrid poplar tend to be planted on river bottoms and require no irrigation. In dry regions, such as east of the Cascades, the poplar are planted on lands formerly in irrigated agricultural crops. In dry regions, drip irrigation is common. Yields are in the fifty cubic meters per hectare range on irrigated sites and thirty to forty cubic meters on nonirrigated sites. In addition to the 20,000 to 25,000 hectares of fiber farms currently in the Pacific Northwest and British Columbia, there are plans for continued increased plantings.[20]

Fiber farms are also being established in the U.S. South and East, motivated by the growing local scarcity of short fiber. Hardwood is increasingly coming into short supply due, in part, to the fact that much of it is located on more inaccessible, wetter sites. Fiber farms are being located especially on deep, well-drained sandy soil, often along rivers. Westvaco, for example, has a cottonwood fiber farm in Missouri, which grows hardwood using fertigation to increase their growth. *Fertigation* is a process that delivers precise amounts of water and fertilizer through a drip irrigation system. The anticipated rotation period is as short as five years (Withrow-Robinson, Hibbs, and Beuter 1995).

The reasons for the emergence in North America of fiber farms appear to be twofold. First, they have been created to meet an anticipated shortage of short fiber, especially near certain pulp mills, over the next decade or two. Second, they are viewed as a possible approach to dealing with the competition expected to come from tropical plantations over the next several decades. Experience from South America, especially in Aracruz in Brazil, has demonstrated that improving the seed quality can dramatically increase growth yields.

CHANGING FOREST AND LAND USE LAWS AND REGULATIONS THAT AFFECT FORESTRY AND FOREST PRACTICES

One of the most important changes affecting forest productivity is the advent of a host of new forest policies, practices, and regulations that have been instituted in a number of countries, literally across the globe. These consist of land set-asides in the form of protected preserves, which preclude timber harvests, as well as changes in policies and logging practices, which reduce the amount of timber that can be removed from a site. Also, changes in various logging and management practices are seen as increasing the costs of logging operations. This section discusses the nature of some of these changes and their impacts on forest harvest productivity.

Recent Trends

Recent decades have seen a burgeoning of protected areas worldwide. Figure 5-5 shows the cumulative areas of all types under protected status since 1900. Although it took until 1960, sixty years, for the first million square kilometers to receive protected status, the second million took only about ten years. By 1990, nearly five million square kilometers were in protected status, about three million square kilometers of which were placed into protected status in the most recent fifteen-year period for which data are available, 1975–1990 (Reid and Miller 1993).

In recent years, the United States has seen substantial changes, which have reduced significantly the harvests from public forests. However, reduced harvests from public lands translate into higher prices for the timber that is harvested from private lands. For years there were disputes over the harvest levels that should be taken from federal forestlands, especially from the heavily forested areas of the U.S. West. Although the harvest-level determinations were ostensibly made by the Forest Service

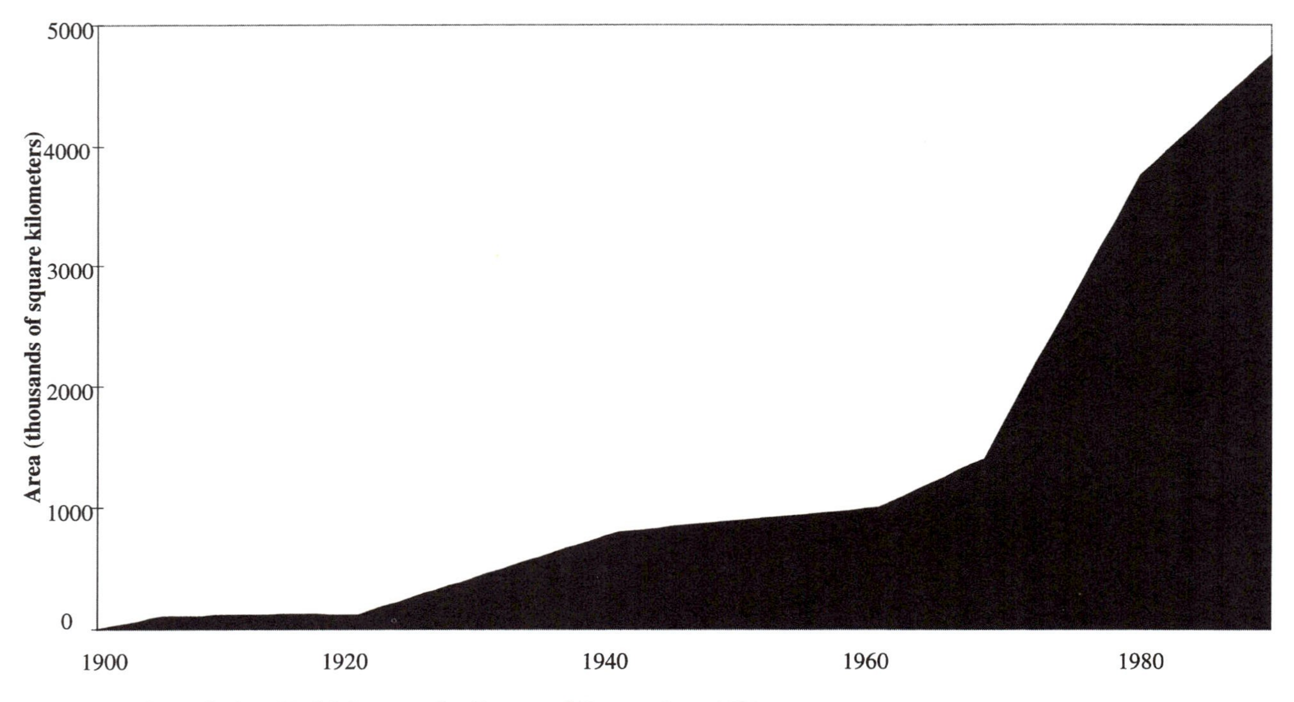

Figure 5-5. Cumulative World Area under Protected Status since 1900.

Source: Reid and Miller 1993.

on the basis of professional considerations, in many years harvest levels were dictated by Congress on the basis of political considerations.

In the latter part of the 1980s and early 1990s, public harvest levels in some regions, especially the Pacific Northwest, were being tied up by court injunctions related to the Endangered Species Act and the spotted owl. Some of those issues were resolved, at least for the time being, by the Forest Ecosystem Management Assessment Team in 1993. In the last couple of years, harvests on federal forests in the United States have fallen substantially, reflecting new political realities.

The 1990s have also seen a tightening of forest practices on private sector forests. First, many states have forest practices codes that are legally binding. These provide standards for logging and forestry within those states. These standards, which include restrictions on riparian zones, roads, and so on, generally raise costs and reduce allowable harvest from any individual site. Additionally, the forest industry in the United States is voluntarily undertaking a Sustainable Forestry Initiative whereby the firms commit to certain forest practices and to meeting certain standards.

Despite these factors, or perhaps because of them, worldwide timber harvests in 1994 were almost identical to harvests ten years earlier in 1984. In the United States, timber harvests have been maintained at close to peak levels with the 1994 harvests being almost 400 million cubic meters, down only slightly from the peak of 427 million in 1990, albeit at higher overall prices. The maintenance of the harvest level despite timberland withdrawals, federal harvest declines, and more stringent forest harvest regulation is due to increased harvests on private lands in response to higher prices. In addition, concerns that future harvest restrictions could become even tighter have accelerated the harvest.[21] Also, some of the U.S. shortfalls relative to domestic demand have been met by increased imports from Canada and from reduced wood exports abroad.

Furthermore, the rate of forest set-asides appears to be increasing. In the United States, harvests from U.S. public lands have decreased dramatically over the past few years as forestlands have been, de facto, withdrawn from the timber base. For example, harvests on U.S. Forest Service land fell in 1995 to less than one-third their 1989 level, with only about 6% of total U.S. production coming from Forest Service lands in the mid-1990s. These declines have little to do with the physical availability of native forest, since forest stocks on federal lands are expanding. Rather, most of these declines are the result of policies that reduced logging on public forests. Furthermore, harvest reductions of this type are not limited to the United States. In Canada, for example, British Columbia's new forest practices code calls for harvest reductions of 5 to 15% overall.

Sustainable Forestry and Certification

Coming out of the United Nations Conference on Environment and Development (UNCED) meeting in Rio de Janeiro in 1992 have been a host of forestry initiatives designed to limit deforestation and/or improve forest management and logging practices so as to reduce environmental damage and maintain traditional forest ecosystems. Included among these are efforts to reduce harvests on native old-growth forests, as discussed above, and to move forestry practices toward what is being called *sustainable forestry*. Sustainable forestry is different from the more traditional concept of sustained-yield forestry in that the concept is designed to "sustain" not only timber yields, but also a host of other forest outputs and ecological services, including such outputs as biodiversity.

The followthrough to the initiative is proceeding on three levels. Most governments in the industrial world have been engaged in one of the ongoing international dialogues. These have included the Helsinki ministerial meeting (involving the European countries), the Montreal process (involving largely non-European industrial countries), and the Santiago Declaration (which was a continuation of the Montreal process). These meetings have been examining ways for the industrial countries to move toward agreements of mutually acceptable approaches toward the management of their forests. In general, the process has been difficult since the forest ownership patterns vary considerably among countries; for instance, almost 90% of Canadian forests are publicly owned, while in the United States about 80% of the forests are privately owned.

At the national level, most countries are modifying their forest policies and forest practices codes to reflect international concerns. Finally, at the private and nongovernmental level, changes are also occurring. Where forestland is privately owned, owners are concerned as to the nature of any new policies that may emerge. Where forestland is publicly owned but the industry privately owned, the industry is concerned as to how policy changes might affect their wood supplies. In many countries, the industry is taking initiatives to modify forest practices. These private actions are also interacting with the activities of some nongovernmental organizations, which have environmentalist agendas calling for formal third-party certification.

In the United States, for example, the major industry association, the American Forest and Paper Association, has established a Sustainable Forestry Initiative, which calls on industry members for a commitment to a minimum standard of forest practices. Although most of the large forest products firms have committed to meeting these standards, the industry also is generally resisting efforts for third-party certification of their forest practices. Additionally, in the United States, almost 60% of the forestland is owned by nonindustrial private forest landowners. These owners, by definition, have no processing capacity and are for the most part small, individual forestland holdings. The willingness of these owners to adhere to the Sustainable Forestry Initiative is uncertain.

Costs of Forest Practices: Some Estimates

Sustainable forestry often involves the introduction of new management techniques. Typically, costs are increased due to the set-aside of larger areas for riparian zones, the retention of some of the trees, and other practices that increase costs. Firms in the United States have commonly estimated the increased costs to be of the order of 5 to 10%. In a recent paper, Greene and others (reported in Michaelis 1996) estimate the reduction in harvest for the Pacific Coast region of the United States due to regulations to be from 5 to 14% and at a cost increase of 7%.

In British Columbia, a host of different studies have been done to estimate the effects of the new forestry code. An early study estimated increased harvest costs at about $220 million annually (van Kooten 1994), while a more recent assessment in British Columbia of the cost of the B.C. Forest Practices Code (Haley 1996) estimates that the Code's implementation would increase total annual costs by about $1.5 billion. In another study using a general equilibrium model, Binkley and others (1994) estimated the annual losses to provincial GNP due to the Forest Practices Code at more than $1 billion annually or 1.54% of British Columbia's annual gross domestic product. Binkley (1995) estimated that the effects of the current policy regime will result in about a 23.5% long-term reduction in total harvest levels in British Columbia.

Globally, environmental pressures are resulting in modified logging and other forest practices throughout much of the world (see Sedjo and others 1998). In the Nordic countries of Finland and Sweden, for example, new forest policies have placed biodiversity on an equal footing with industrial wood production. The result has been modified logging practices that, among other changes, reduce the area eligible for harvest and leave more standing trees on harvested areas. For many situations, the new harvest policy is estimated to increase costs 5% to 10%. However, a recent study by Kajanus and Karjalainen (1996) for the Nordic countries found that the costs can rise as much as 30 or 40% for certain land conditions. This was largely because the practices required to achieve third-party certification would reduce harvests due to set-asides. Also, costs would rise due to restoration targets and the alteration of certain management methods. They estimated that the average net effect would be to

Table 5-4. Some Cost Estimates of Increased Environmental Standards.

Country	*Study*	*Cost Increase*
United States	Greene and others 1995	5–14% cost increase
Finland	Tikkanen 1997	12% decrease in net income to meet new biodiversity standards
Finland	Kajanus and Karjalainen 1996	18% decreased net income to meet certification standards
Canada	van Kooten 1994	$3 cost increase/cubic meter
Canada	Haley 1996	$8 cost increase/cubic meter (about 15% cost increase) + other costs

reduce income per hectare by about 18%. Some estimates are presented in Table 5-4.

IMPLICATIONS AND CONCLUSIONS

Implications for North American Comparative Advantage

North America has held a comparative advantage in industrial wood production based largely on the existence of a huge native forest covering much of the United States and Canada. The stock of United States old-growth is essentially gone, with most of the remaining old-growth forest having been set aside in protected areas. The United States will need to find other sources of industrial wood and will surely find an increasing share of its production coming from intensively managed forests. The United States is likely to continue to move into intensive management land areas that are well-suited to plantation forestry. These areas, mostly in the South, have relatively high biological growth rates, are very accessible, are well located vis-à-vis markets, have relatively flat terrain, and present few environmental problems.

Canada will be able to maintain its position as a global wood producer by virtue of its ability to continue to draw from its remaining large areas of native forest. It has been able to maintain its share of the U.S. wood market, as well to export widely outside of North America, by harvesting from its flat, accessible areas of native forest. In the difficult coastal area of British Columbia, however, the forest harvest has been maintained at more or less the same levels for a considerable period because it has a unique market, that is, the high-price Japanese wood market.

While many parts of the United States have climate, soils, and so on suitable to effective intensive forestry, the United States has no monopoly on favorable timber-growing conditions and its plantations. Although competitive, the U.S. plantations are unlikely to be the low-cost producer worldwide.

Worldwide, some regions have already make great strides in developing a competitive, intensively managed, forest industry.[22] However, not all of the regions that have the natural prerequisites for industrial forestry have the necessary political, social, and cultural prerequisites for successfully developing a sustainable forest production industry.[23]

Conclusions

Although traditionally most of the nation's industrial wood came from the harvest of natural forests, humans possess the knowledge and expertise to create forests for the production of industrial wood and/or other outputs. Unlike other natural resources examined in this suite of RFF productivity studies, the forest resource is renewable. In recent decades, investments in planted forests have been expanding rapidly, and humans are well into a transition away from reliance on natural forests to utilizing wood produced through conscious human investments.

This study concludes that the dominant innovation within the industry is the move to intensive plantation management. There are two reasons for this. First, obtaining wood from the extensive margin (that is, from old-growth forest) is becoming increasingly difficult and uncertain as old-growth forests are becoming increasingly difficult and costly to harvest, due to inaccessibility and to politically generated forest set-asides. Second, the opportunities at the intensive margin, such as intensively managed forest plantations, have been demonstrated to be especially favorable on high biological sites in locations with access to processing facilities and to markets, both at home and abroad, due to the ability of intensive forest management to increase yields to several times those of nonmanaged forests.

The study also concludes that there has been no recent, single, dominating new technology in logging and that productivity levels have, at best, grown modestly. Innovations are barely keeping pace with cost-increasing factors in harvesting, including poorer accessibility of old-growth forests.

The trend to intensively managed forests is driven by technological changes that include increased knowledge of tree-growing potential, including genetic improvements in tree growing. The move to forest planting draws heavily from the techniques developed in agriculture cropping, with appropriate modifications. Forestry is now focusing on utilizing genetic techniques, many developed earlier in agriculture, to upgrade their seedlings and genetic stock.

This study also examined the effects on the timber industry of public policy that imposes more stringent and expensive harvesting practices and also withdraws timberlands from the industrial forest land base. This analysis suggested that timberland set-asides will increase the value of the remaining timberlands and that the increased harvesting costs will provide incentives to invest in intensively managed plantation forests. These considerations make intensive management more attractive.

In summary, due to both technological opportunities and as a reaction to public policy, industrial wood production appears to be shifting to intensively managed forest plantations and fiber farms where returns from genetically improved growing stock can be captured by the producer. The transition to intensive wood management has been underway for a few decades, and the process has now become well developed. Plantation management now accounts for an estimated one-third of global industrial wood production, with perhaps 15% attributable to intensively managed plantations. However, the potential for further innovation and expansion is vast and the process has just begun.

ACKNOWLEDGMENTS

I would like to acknowledge the valuable assistance of many individuals and organizations. Field trips to relevant operations included: a logging operation near Joensuu, Finland, and discussions with Bjorne Hagglund and correspondence with Hans Troedsson, both of Stora Skog, Sweden; a logging training operation on the Olympic Peninsula and conversations with Peter Schiess, University of Washington; visits and discussions with employees of MacMillan Bloedel, Vancouver, British Columbia, including Bill Caferata, Neil Brett-Davies, Rob Prinz, Bill Beese, Brett Gildner, G.B. Dunsworth, S.M. Northway, and N.J. Smith. I also had helpful discussions with almost a dozen members of the Forest Engineering Research Institute of Canada, Vancouver, B.C., led by A.W.J. Sinclair and including Ernst T. Stjernberg, Marv Clark, Raymond K. Krag, and Ingrid B. Henin. Clark Binkley, dean of the faculty of forestry, University of British Columbia, also provided useful comments.

Information regarding logging in the U.S. South was provided by William Stuart and William Hyde of Virginia Polytechnic Institute and State University, Fred Cubbage of NCS, and Cam Carte of the AF&PA. Visits and discussions on plantation forestry and forest genetics were undertaken with Robert Rogers and others at MacMillan Bloedel and with Rex McCullough and Chris Dean of Weyerhauser Co., Tacoma, Washington. A field trip of hybrid poplar fiber plots was undertaken with Toby Bradshaw, a University of Washington molecular geneticist. William Baughman of Westvaco, Summerville, South Carolina, and his staff—including Brian Fiacco, Gene Kodama, Ed Owens, John Thurmes, Leland Gauron, Dave Gerhardt, Dave Canavera, Dick Dainiels, and Cliffort Schneider—provided much useful information. David Duncan and Lisa Drake of Monsanto provided information on genetics and pesticide/herbicide use.

Finally, I would like to thank Neil Brett-Davies, Sharon Friedman, William Hyde, Cliff Schneider, and William Stuart for their comments on earlier drafts of this chapter.

ENDNOTES

1. In practice, growth yields are typically increased two to four times.

2. Until recently tropical hardwoods were used rarely in paper production. However, recent technological innovations allow for the use of mixed tropical hardwood in the production of certain papers. Modern technology is allowing a large degree of interspecies substitutability.

3. Today, almost 40% of U.S. consumption of softwood lumber is imported from Canada. One and two decades ago, it was below 30%.

4. Deforestation in the tropics is driven largely by land conversion to nonforest uses, primarily agricultural uses (UN FAO 1987).

5. Not all wood is identical for industrial wood purposes. Where forests are very heterogeneous, as in many tropical forests, most of the species do not have ready markets.

6. I am indebted to William P. Stuart of Virginia Polytechnic Institute and State University for enlightening me as to the role social forces have played in the development of technologies in the various regions, and I have borrowed heavily from his suggestions.

7. Ruttan (1982, 237) gives the following definitions for technical change: Technical change (a) is the substitution of "inexpensive and abundant resources for scarce and expensive resources," (b) is "the substitution of knowledge for resources," and (c) "releases the constraints on growth imposed by inelastic resources supplies."

8. Nordic mills have provisions for the buildup of very large wood inventories.

9. With a constant quality and accessibility of the resource, technological improvements would be reflected in declining delivered resource prices. Where the quality/accessibility of the resource is declining, improved technology must first offset the higher real costs of the delivery.

10. Access to old growth in the Pacific Northwest region has become more difficult as logging moved from the flat coastal plain into the rugged interior mountains.

11. Quality attributes include log size, absence of knots, species type, and so forth. Sawlog size in the South, for example, had declined steadily during the post-WWII period. Sedjo and Lyon (1990) suggested that declines in wood quality have been offset by technological improvement in processing.

12. Farnum, Timmis, and Kulp (1983) showed that theoretical maximum biological timber yields were five to ten times natural yields, although in practice the growth increase is usually two to four times.

13. Pinus radiata, know as Monterey pine in its native California, grows much more rapidly in Chile and New Zealand. Also, the U.S. Southern pines grow more rapidly in southern Brazil and Argentina than in the U.S. South.

14. For a discussion of the advantage of industrial plantations, see Sedjo 1983.

15. This does not treat as plantation production the wood of the U.S. South or most of Europe.

16. Much of the information reported in this section about fiber farms was provided by Toby Bradshaw of the University of Washington (Bradshaw 1997).

17. Growth in hybrid poplar stands is five to ten times the rate experienced in native forest growth rates (Bradshaw 1997).

18. In addition to being a very effective herbicide, glyphosate has the desirable properties of decomposing quickly on the site and so avoiding residual and downstream damages (Monsanto, no date, and personal conversation with David Duncan, Monsanto Company).

19. It is estimated that four to five million trees are planted in the United States *every day.*

20. For example, MacMillan Bloedel is planning to plant about 1,000 hectares annually beginning in 1997 (personal communication Neil Brett-Davies; memo from S.L. Tisdale 6/18/97).

21. It is common to hear of private owners who rush to harvest in areas of unoccupied spotted-owl habitat due to concerns that, if not harvested, spotted owls may find nesting within the area thus precluding future harvest.

22. Brazil, for example, reversed a $500 million pulp and paper trade deficit into an almost $1.5 billion trade surplus in less than twenty years (Sedjo 1998).

23. Although many regions may not have the political prerequisites to allow for long-term investments as required in forestry, many regions obviously do, since the role of the nontraditional producer in tropical and subtropical regions is growing.

REFERENCES

Binkley, Clark S. 1995. Designating an Effective Forest Research Strategy for Canada. *Forestry Chronicle* 71: 589–95.

Binkley, Clark S., Michael Percy, William A. Thompson, and Ilan B. Vertinsky. 1994. A General Equilibrium Analysis of the Economic Impact of a Reduction in Harvest Levels in British Columbia. *Forestry Chronicle* 70(4): 449–54.

Bradshaw, Toby. 1997. Director, Poplar Molecular Genetic Cooperative, University of Washington, Seattle. Personal communication with the author.

Cubbage, Frederick, and Douglas Carter. 1994. Productivity and Cost Changes in Southern Pulpwood Harvesting, 1979 to 1987. *Southern Journal of Applied Forestry* 18(2, May).

Farnum, P., R. Timmis, and J.L. Kulp. 1983. The Biotechnology of Forest Yield. *Science* 219: 694–702.

Greene, John L., and others. 1995. The Status and Impact of State and Local Regulation on Private Timber Supply. General Technical Report RM-255. Ogden, Utah: USDA Forest Service.

Haines, R.J. 1994a. Biotechnology in Forest Tree Improvement. FAO Forestry Paper 118. Rome: United Nations Food and Agricultural Organization.

———. 1994b. Biotechnology in Forest Tree Improvement: Providing Technology for Wise Use. *Unasylva*, 45(177).

Haley, David. 1996. Paying the Piper: The Cost of the British Columbia Forest Practices Code. Paper presented at the Working with the B.C. Forest Practices Code and Insight Information Inc. Conference, sponsored by the *Globe and Mail*, Vancouver, British Columbia, April 15–17.

Hayami, Yujiro, and Vernon W. Ruttan. 1971. *Agricultural Development: An International Perspective.* Baltimore: Johns Hopkins University Press.

Hyde, W.F., D.H. Newman, and Berry J. Seldon. 1992. *The Economic Benefits of Forestry Research.* Ames, Iowa: Iowa State University Press.

Kajanus, Miika, and Harri Karjalainen. 1996. WWF's Ecolabelling Project: Costs of Ecolabelled Forestry. Paper presented to the meeting of the Scandinavian Forest Economics, Finland, on 20 March.

Laestadius, Lars. 1990. A Comparative Analysis of Wood-Supply Systems from a Cross-Cultural Perspective. PhD Dissertation, Virginia Polytechnic Institute and State University, Blacksburg, Virginia. July.

Michaelis, Lynn O. 1996. Challenges to Private Land Management in the Pacific Northwest: Is There a Future? Paper presented to Conference on Forest Policy: Ready for Renaissance, Olympic Natural Resource Center, Forks, Washington, September.

Monsanto. No date. *Ecological and Environmental Aspects of Monsanto Forestry Herbicides.* St. Louis: Monsanto Agricultural Company.

Newman, D.H. 1987. An Econometric Analysis of the Southern Softwood Stumpage Market: 1950–1980. *Forest Science* 33(4): 932–45.

Parry, Ian W.H. 1997. Productivity Growth in the Natural Resource Industries, 1970–1994. RFF Discussion Paper 97-39. Washington, D.C.: Resources for the Future.

Reid, Walter V., and Kenton B. Miller. 1993. *Keeping Options Alive: the Scientific Basis for Conserving Biodiversity.* Washington, D.C.: World Resources Institute.

Ruttan, Vernon W. 1982. *Agricultural Research Policy.* Minneapolis: University of Minnesota Press.

Sedjo, Roger A. 1983. *The Comparative Economics of Plantation Forestry: A Global Assessment.* Washington, D.C.: Resources for the Future.

———. 1997. The Forest Sector: Important Innovations. RFF Discussion Paper 97-42. Washington, D.C.: Resources for the Future.

———. 1998. The Potential of High Yield Plantation Forestry for Meeting Timber Needs: Recent Performance and Future Potentials. In *Planted Forests.* Boston: Kluwer.

———, and Kenneth S. Lyon. 1990. *The Long-Term Adequacy of World Timber Supply.* Washington, D.C.: Resources for the Future.

———, Alberto Goetzl, and Steverson O. Moffat. 1998. *Sustainability of Temperate Forests.* Washington, D.C.: Resources for the Future.

Solow, Robert. 1974. The Economics of Resources or the Resources of Economics. *American Economic Review* 64(2): 1–14.

Stier, Jeffrey C. 1982. Changes in the Technology of Harvesting Timber in the United States: Some Implications for Labour. *Agricultural Systems,* 255–66.

——— and David N. Bengston. 1992. Technical Change in the North American Forestry Sector: A Review. *Forest Science* 38(1, February).

Tikkanen, I. 1997. Personal conversation, March 12, 1997.

UN FAO (United Nations Food and Agricultural Organization). 1987. *Tropical Deforestation.* Rome: UN FAO.

———. 1995. *Forest Yearbook.* Rome: UN FAO.

van Kooten, G.C. 1994. Cost-Benefit Analysis of B.C.'s Proposed Forest Practices Code. Vancouver, British Columbia: Forest Economics and Policy Analysis Research Unit of British Columbia.

Westvaco. 1996. Fiber Farm Expansion. *Forest Focus* 20(4, March).

———. 1997. *Forest Focus* 21(1, March).

Withrow-Robinson, Brad, David Hibbs, and John Beuter. 1995. Poplar Chip Production for Willamette Valley Grass Seed Sites. Research Contribution 11, College of Forestry, Forest Research Laboratory, Oregon State University.

6

Productivity Trends in the Natural Resource Industries

A Cross-Cutting Analysis

Ian W.H. Parry

People have long been concerned about the exhaustible nature of natural resources and the possible limits on economic growth caused by resource scarcity. In the early nineteenth century, Thomas Malthus predicted that finite quantities of land and other types of capital would constrain population growth and improvements in living standards. David Ricardo argued that the costs of producing output would rise over time as the higher-quality deposits of natural resources were used up.

Contrary to these gloomy predictions, most natural resources remained abundant throughout the nineteenth century and the first half of the twentieth century. The United States and Europe experienced rapid population growth and economic development. Thus, economists paid little attention to the problems of resource scarcity. During the 1940s, however, the prices of agricultural and other natural resource products rose. This stimulated more research on natural resource industries, including the seminal study by Barnett and Morse (1963). They found that average production costs and prices had been falling in the natural resource industries in the United States during the period 1870 to 1957. Indeed, the price increases of the 1940s were at least partially reversed in the 1950s. Barnett and Morse concluded that the effective resource base had been increasing over time. That is, new discoveries of deposits and the ability to make use of lower-grade ores because of technological advances had more than compensated for the depletion of resource stocks.

IAN W.H. PARRY is a fellow at Resources for the Future.

The post-World War II economic boom came to an abrupt end in the 1970s. The level of productivity for the U.S. economy as a whole actually fell.[1] At the same time, the prices of agricultural land and most natural resources increased markedly. This led to studies claiming that the world was running out of resources and that the limits to economic growth were being reached.[2] However, the most prominent studies contained some methodological problems and their predictions turned out to be well wide of the mark.[3] Indeed, since the 1980s economic growth and productivity growth for the whole economy have recovered, although not quite to the rates experienced between 1945 and 1970.

This study attempts to update the work of Barnett and Morse by looking at economic performance in a number of natural resource industries since 1970. We focus on productivity rather than average production costs. This is because the latter measure is directly affected by changes in input prices (for example energy prices), and these are typically caused by developments in other markets. We present a "top down" statistical analysis of productivity trends in four representative natural resource industries: coal, petroleum, copper, and logging. Other chapters in this book by Darmstadter, Bohi, Tilton and Landsberg, and Sedjo present "bottom up" analyses of these four industries. That is, these other studies discuss in detail changes in the state of technology, industry structure, the regulatory environment, and so on that have affected productivity levels during the period.[4]

Our objective in these studies is to focus as much as possible on the *extraction* of natural resources rather than the *processing* of natural resources into products, since the former is more important for resource availability. Previous work in this area has tended to use broader industry definitions that incorporate downstream activities.[5] We use a multifactor definition of productivity, which is more reliable than simple labor productivity. The analysis follows the growth accounting method, which decomposes changes in industry output into changes in the quantity of inputs and a residual that reflects changes in productivity.

Productivity performance in the four industries was much more erratic over the last twenty-five years, compared to the steadily increasing productivity common to each industry prior to 1970. All four industries, and particularly petroleum, experienced significant negative growth in measured multifactor productivity during the 1970s. However, in retrospect, the 1970s look like an exceptional period, rather than marking a change in long-run productivity trends. The productivity declines of that decade appear to be explained by a number of special factors that generally have a transitory rather than a permanent effect on productivity growth. For example, the rise in natural resource prices encouraged the entry of relatively inefficient producers. New environmental and health and safety regulations phased in during the period also reduced productivity measures.

Since the early 1980s, productivity growth has resumed in all the industries, although by markedly different amounts. For example, we estimate that the level of productivity in 1992 was around 75% higher in the petroleum industry than at the trough of the productivity slowdown and around 60% higher in coal and copper. To some extent, the productivity improvements represent restructuring and consolidation in response to falling output prices. However, technological developments have also played an important role in all four industries.

Thus the productivity performance of the industries analyzed below, and by implication resource scarcity in the United States, appears to be much less of a worry today than in the 1970s. There are some caveats to this comforting conclusion. First, the rate of productivity improvements in recent years may not necessarily continue in the future. Indeed, to the extent that the recent improvements represent a once-and-for-all restructuring in response to a more competitive global environment, we might expect productivity growth to slow down somewhat in the future. Second, not all natural resource industries are represented in our study. In particular, we do not consider the fishing industry, where excessive depletion can occur because of the open-access nature of many fisheries.[6]

In this chapter, we first describe the theoretical framework for deriving a formula for multifactor productivity and then discuss how measures of output and input quantities in the four industries are obtained. Next we present our estimates of productivity trends in each industry, along with other measures of industry performance and finally offer a brief summary.

THEORETICAL FRAMEWORK

The theoretical analysis below is based on the growth accounting framework developed primarily by Denison, Kendrick, Jorgenson, and Griliches.[7] We begin by specifying the following production function:

$$Q_t = A_t F(L_t, K_t, M_t) \tag{6-1}$$

Q_t is physical output in an industry (tons of copper, and so on) at time t, and $F(.)$ is a function. L_t is labor input. This is the total number of hours worked, which in turn are the average hours per employee, multiplied by the total number of employees. K_t is the flow of services from the physical capital stock, broadly defined to include buildings and structures, land, equipment, and inventories. The measurement of K_t is discussed below.

M_t is intermediate goods input. This consists of raw materials, purchased services, and energy.[8]

A_t is a residual scalar, representing all unobserved factors that may cause the production function to shift over time. There are four potentially important determinants of changes in A_t. Improvements in the state of technology comprise the first determinant. These include more efficient techniques for combining inputs in production (disembodied technological change) and improvements in the quality of the capital stock, such as better machines (embodied technological change). Changes in economies of scale arising from the entry and exit of firms into the industry when these firms differ in efficiency from incumbent firms comprise the second determinant. As the third determinant, regulations can affect the level of productivity by diverting inputs away from producing output to other activities, such as improving health and safety conditions in the workplace or the surrounding environment.

Each of these three factors affects "true" productivity. However, a fourth factor—changes in the quality of exploited resources—also shows up in our productivity estimates. Ideally, we should treat the natural resource stock as an input in the production function. In this case, as the industry moves to lower-quality resources over time, this would be represented by a decline in the resource input. However, this approach would be difficult to implement because the amount of quality-adjusted natural resource input is very difficult to quantify. What this means is that our measure of productivity changes is less than true productivity changes (that is, changes in output less changes in labor, capital, intermediate goods, and resource inputs) to the extent that the average quality of the remaining natural resource is falling over time.[9]

Firms are assumed to maximize profits by taking the price of output (p_t), labor (w_t), capital (r_t), and intermediate goods (v_t) at time t as given. This price-taking assumption seems a reasonable approximation, given the large number of firms in each of the four industries and that no one firm produces a substantial share of industry output.[10] Therefore we can use the familiar first-order conditions, equating value of marginal products with factor prices:

$$p_t \frac{\partial F}{\partial L_t} = w_t; \quad p_t \frac{\partial F}{\partial K_t} = r_t; \quad p_t \frac{\partial F}{\partial M_t} = v_t \tag{6-2}$$

where w_t, r_t, and v_t denote the price of labor, capital, and intermediate goods, respectively. Differentiating Eq. 6-1 with respect to time gives:

$$\dot{Q}_t = \frac{\partial F}{\partial L_t}\dot{L}_t + \frac{\partial F}{\partial K_t}\dot{K}_t + \frac{\partial F}{\partial M_t}\dot{M}_t + \dot{A}_t F \tag{6-3}$$

where $\dot{L} = dL/dt$, and so on. Substituting Eq. 6-2 in Eq. 6-3 and dividing by Q gives:

$$q_t - \{\pi_t^L l_t + \pi_t^K k_t + \pi_t^M m_t\} = a_t \tag{6-4}$$

where lower case letters denote growth rates ($l_t = \dot{L}/L_t$, and so on), and

$$\pi_t^L = \frac{w_t L_t}{p_t Q_t}; \quad \pi_t^K = \frac{r_t K_t}{p_t Q_t}; \quad \pi_t^M = \frac{v_t M_t}{p_t Q_t} \tag{6-5}$$

π_t^L, π_t^K, and π_t^M are the shares of payments on labor, capital, and intermediate inputs in value product.[11]

In Eq. 6-4, a_t is the growth rate of multifactor (or total factor) productivity at a point in time. This is equal to the growth rate in output less a weighted combination of the growth rate in the quantity of labor, capital, and intermediate inputs. Many other studies focus on labor productivity rather than a multifactor measure of productivity. However, in the industries analyzed here, capital and intermediate goods are important inputs in production. Using a simple labor productivity measure may give a misleading impression of overall productivity (in quantitative, though not necessarily qualitative, terms): labor productivity growth overstates multifactor productivity growth when the quantity of capital and intermediate inputs are increasing relative to labor input over time.

Some studies (for example those by Denison and by Jorgenson and Griliches) use a productivity measure that is net of increases in input quality, while others (for example, Kendrick) use a gross productivity measure. Due to data limitations at our level of industry disaggregation, measures of input quality are necessarily crude. For this reason we do not attempt to adjust inputs for changes in quality.[12]

Finally, since data are not available on a continuous time basis, we replace the continuous growth rate variables in Eq. 6-4 by the analogous discrete time growth rates:

$$l = (L_{t+1} - L_t)/L_t \tag{6-7}$$

and so on.

THE MEASUREMENT OF INDUSTRY OUTPUT AND FACTOR INPUTS

This section discusses various issues in the measurement of output and input series for each of the industries. The data sources are then described.[13]

Industry Output

Stage of Production. Output for each industry is defined at the extraction level from production on both public and private lands. For petroleum (oil and natural gas), this industry definition captures drilling activities from on- and offshore wells, but not refining into gasoline, and so forth. Similarly, coal and copper output is volume produced from the mines. For logging, output is after the trees have been harvested and cut into logs.[14] In each case, measured inputs do not include those used to transport the commodity to processing facilities. The SIC codes are 1311 (petroleum), 122 (coal), 1021 (copper), and 2411 (logging).

Changes in the Quality of Output. Ideally, to compare output at different points in time, allowance should be made for changes in the quality of market output over time. However, for the industries studied here, changes in the quality of market output have probably been negligible over the period, because they essentially produce unprocessed raw materials. For example, the quality of a barrel of oil or natural gas, measured by Btu content, is more or less the same now as it was twenty-five years ago. We make no adjustments for changes in quality.[15]

Changes in the Composition of Output. The composition of industry output may change over time, for example the ratio of natural gas to total oil output increased by 20% between 1970 and 1994. However, in terms of pure energy use, a Btu of oil is identical to a Btu of natural gas. Therefore, petroleum output is measured by the sum of oil and natural gas, measured in terms of Btus.[16] The average Btu content of a ton of coal has declined by 10.5% over the last twenty-five years (due to the substitution in favor of low-sulfur content coal). To incorporate this effect, coal output is measured by the volume of coal output multiplied by the average Btu content. In the copper industry, production is measured by total tonnage. Finally, logging output is roundwood, which comprises the sum in cubic meters of hardwood, softwood, and (less important) scrap wood. We do not make an adjustment for changes in the composition of logging output, since these have only been very slight over the last twenty-five years.

Gross Output or Value Added? Industry output could be defined as gross (as above) or net of intermediate goods purchased from other industries. We use the more general gross output definition, since this enables us to include an assessment of the contribution of intermediate inputs relative to labor and capital in industry output growth. In addition, only value added can be measured, and this is an imperfect measure of output added, since it is affected by changes in the product price.

Output Gross or Net of Depreciation? Output can be defined as gross or net of depreciation of the industry's capital stock. The latter definition gives a more accurate indication of the welfare to society from industrial output. This is because welfare depends on what is available for current consumption and net additions to the capital stock, and not on the amount of resources used up in maintaining existing capital goods. However, in this chapter, we are attempting to infer the state of technology in the industries—that is, the ability to transform inputs into outputs. Since technological changes affect output *gross* of depreciation, this is the appropriate measure to use (Hulten 1992).

Nonmarket Impacts. There are three types of nonmarket impacts caused by the extraction of natural resources:

1. environmental impacts, such as the loss of natural habitat from logging and damage caused by oil spills;
2. possible adverse health effects for workers within the industry, for example, black lung disease caused by coal mine dust; and
3. changes in the stocks of natural resources. These stock changes could be positive or negative, depending on whether depletion is more than offset by new discoveries and technical improvements enabling the use of lower-grade ores.

Again, if we were examining productivity from a social welfare perspective, it would be appropriate to use a broader measure of industry output that took into account these nonmarket impacts.[17] However, our study is attempting to infer changes in the state of technology in each industry for transforming inputs into market output, and for this purpose it is appropriate to use the conventional measure of industry output.

Gross or Net of Inventories? Finally, output is measured by production rather than sales; that is, it includes additions to inventory stocks.

Labor Input

It is straightforward to obtain an aggregate measure of the quantity of labor because data are readily available on hours worked in the four industries. These are actual rather than potential hours; that is, they do not include time lost because of strikes, sick days, and vacations.[18]

Capital Input

Capital inputs are decomposed into four broad categories in the data: equipment, buildings and structures (henceforth structures), land, and inventories. We first discuss how we obtain measures of the stock of each of these types of capital and then how these are aggregated into a single measure of capital input.

Beginning with the easiest first, data are available on the physical stock of inventories in each period for each industry. Data are not directly available on the physical quantity of land used in the industries. However, the Bureau of Labor Statistics provides estimates of the (constant dollar) value of land holdings in the industries. These are based on the same perpetual inventory method described below, where the depreciation of land is taken to be zero.

The stocks of productive equipment and structures in each period are not directly observed. Information is available on the *book value* of assets—the historical cost of an asset adjusted by some assumption about depreciation—but this can be a misleading indicator of the value of capital.[19] Instead, we adopt the commonly used perpetual inventory method to derive a proxy for these capital stocks. This requires information on gross investment in each period in current dollars, an initial quantity for the particular asset, and an assumption about economic depreciation of each asset over time.

First, we divide the current dollar amount of gross investment in structures and equipment by the respective producer price indexes for structures and equipment in that industry.[20] This gives the constant dollar quantity of investment for each asset. We assume that fraction δ_i of a particular capital stock depreciates each period; thus

$$K_{it} = (1 - \delta_i)K_{i,t-1} + I_{it} \tag{6-8}$$

This equation says that K_{it}, the stock of asset i in period t, equals $(1 - \delta_i)$ times the stock in the previous period plus the current amount of gross investment, I_{it}. By recursive substitution:

$$K_{it} = (1-\delta_i)^t K_{i0} + \sum_{j=0}^{t-1}(1-\delta_i)^j I_{it-j} \tag{6-9}$$

that is, the current capital stock is a weighted average of past investment and the capital stock in the initial period. We obtained estimates of δ_i for equipment and for structures from the Bureau of Labor Statistics.[21] K_{i0} can only be measured by book value divided by asset price. To substantially reduce the influence of this on our capital stock series, we set K_{i0} equal to the capital stock in 1945 for petroleum and coal, twenty-five years before the start of our period of interest.[22]

We now need to attach weights to the individual asset stocks in order to aggregate them into a single measure of capital input. The usual procedure is to weight an individual capital asset by its share of earnings in the total earnings for capital (for each period). Data are not available on capital earnings; hence this has to be estimated. We do this by assuming an earned rate of return on capital. The (ex post) rental rate (r_{it}) on a capital asset is given by:

$$r_{it} = s_t + \delta_{it} - \frac{\dot{p}_{it}}{p_{it}} \tag{6-10}$$

This consists of three components: first, the market rate of interest (s_t), that is, the opportunity cost of using capital rather than lending it out to other users[23]; second, the cost of depreciation of that unit; and third, subtracted from these costs is the rate of (real) price appreciation from holding the asset. The overall productivity statistics are not very sensitive to different assumptions about the values of these components.

To add up different types of capital in an industry at time t, we follow the usual procedure of weighting each asset by its share in the total cost of assets. That is,

$$K_t = \sum_i \lambda_{it} K_{it}; \quad \lambda_{it} = \frac{r_{it} K_{it}}{\sum_i r_{it} K_{it}} \tag{6-11}$$

where K is the aggregate capital stock and λ_{it} is the share of asset i in the total value of capital assets at time t.

Unfortunately, only data on labor input are available for the copper industry, and not expenditures on investment or intermediate goods. However, copper is a substantial share—around 40% during 1970–1994—of the metal-mining industry as a whole, for which these data are available. We assume that the ratio of investment in individual capital assets to nonlabor costs was the same for copper as in metal mining as a whole. Therefore, a proxy for investment expenditures on individual copper assets was obtained by multiplying the relevant series for metal mining by the share of nonlabor costs in copper (that is, the value of output less labor earnings) in total metal-mining nonlabor costs.[24]

Intermediate Goods Inputs

Data are available for expenditures on energy, purchased services, and raw materials, and these need to be converted into constant dollars to obtain an input quantity series. There are separate producer price indexes for energy for each industry, and there is another one for purchased services and raw materials combined.[25] We divide the expenditure series for energy by the price index for energy, and the expenditure series for purchased services and raw materials combined by the other price index, to obtain constant dollar series. These two series are then aggregated using weights equal to the share of the variable in total intermediate goods expenditures.

Again, copper expenditures on purchased services, energy, and raw materials are proxied by those for metal mining as a whole, multiplied by the share of nonlabor costs of copper in metal-mining nonlabor costs. Finally, in logging, we exclude payments for stumpage from intermediate goods. This is because these payments fluctuate substantially, and this mainly reflects changes in rents rather than changes in input quantities.

Aggregating Labor, Capital, and Intermediate Goods

So far we have explained the methodology for obtaining the series for Q, L, K, and M. These are easily converted to q, l, k, and m in Eq. 6-4 by using the rate of growth formula. In order to derive the productivity series, the last step is to obtain the weights attached to each of these three input series, or π_t parameters. These weights are the share of earnings of that input in total value product. Labor earnings and expenditures on intermediate inputs (the sum of expenditures on energy, purchased services, and raw materials)—and hence the weights for these inputs—are easily obtained for each industry. In other industry studies, total capital earnings are often estimated as a residual by subtracting payments to labor and on intermediate goods from the value of output. This is valid in industries that are characterized by constant returns to scale, since the sum of the factors' shares in Eq. 6-4 would be unity (by Euler's theorem). However, in natural resource industries, part of the earnings are rents to the resource stock (thus the sum of the shares is less than unity): using this procedure would count these rents as earnings to capital and overweight the capital share.[26] Therefore, instead, we obtain capital earnings in period t by multiplying the quantity of each capital asset K_{it} by the rental cost of capital (r_{it}) and then aggregating over all assets.[27]

Finally, the value of output is equal to price multiplied by production. The coal and copper prices used are the average price per ton at the minehead. Petroleum prices are a weighted average of the Btu price for oil and natural gas, where the weights are the respective shares in the total value of Btu petroleum output. The total value of production from logging is obtained from published data.

Data Sources

Coal. All the data we used are from the U.S. Department of Energy's *Annual Energy Review*, except investment and expenditure on intermediate goods, which are from the *Census of Minerals Industries*. Intermediate goods are available every four or five years. To interpolate the intervening years, we assume that the share of intermediate goods expenditures in value product changes linearly between the observation points. That is, if the first and fifth year shares are 0.450 and 0.454, then the second year is taken to be 0.451, the third 0.452, and so on.

Petroleum. Production and value product are obtained from the *Annual Energy Review*. Hours worked are from the *Basic Petroleum Data Book*. Investment and expenditure on intermediate goods are from the *Census of Minerals Industries* and, again, the missing years from intermediate goods are interpolated by the same procedure used for coal.

Copper. Output and value product are obtained from the *Minerals Yearbook*. Labor hours and earnings are from *Employment and Earnings*, published by the U.S. Bureau of Labor Statistics. Investment and expenditures on intermediate goods from metal mining (which are scaled down to proxy for copper) are from the *Census of Minerals Industries*, with the same interpolations for intermediate goods.

Logging. Output is from the *FAO Yearbook* and all other statistics are from the *Census of Manufacturers*.[28]

ECONOMIC INDICATORS OF INDUSTRY PERFORMANCE

We now present and discuss statistics on economic performance in the four industries. The first two sections below describe trends in output, price, and share of industry output in total manufacturing output since 1945. The next section discusses labor productivity trends and compares these with trends in broader sectors of the economy. The following two sections describe, and offer some explanations for, our estimates of multifactor productivity trends over the last twenty-five years. The final section presents some summary statistics.

Output and Price Trends

Figures 6-1 and 6-2 show real output and real price—that is, output price relative to the consumer price index—trends for each industry over the last fifty years.[29]

1945–1970. Output in the petroleum and copper industries was on an upward trend from 1945 to 1970, mirroring the steady expansion in the

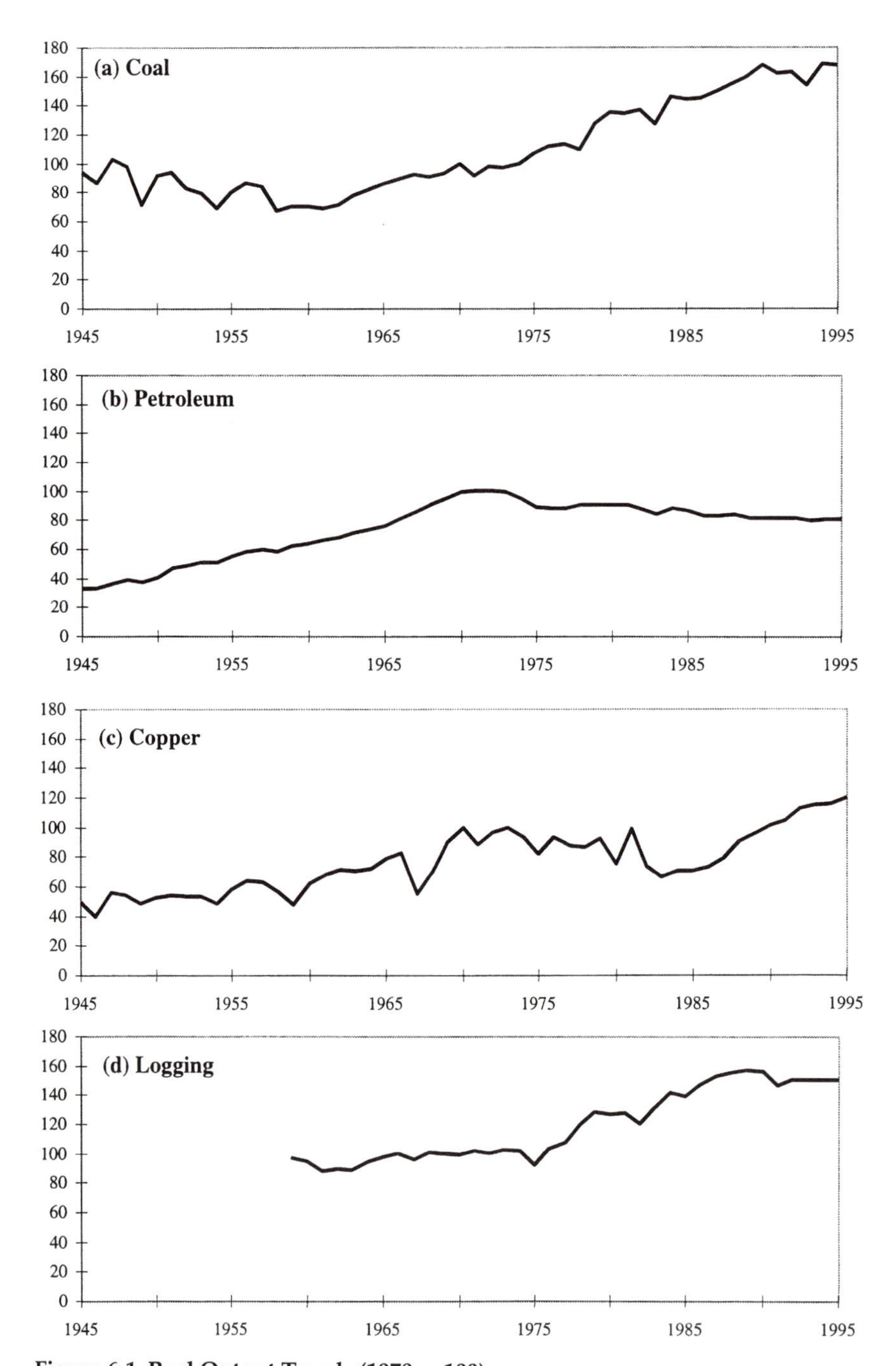

Figure 6-1. Real Output Trends (1970 = 100).

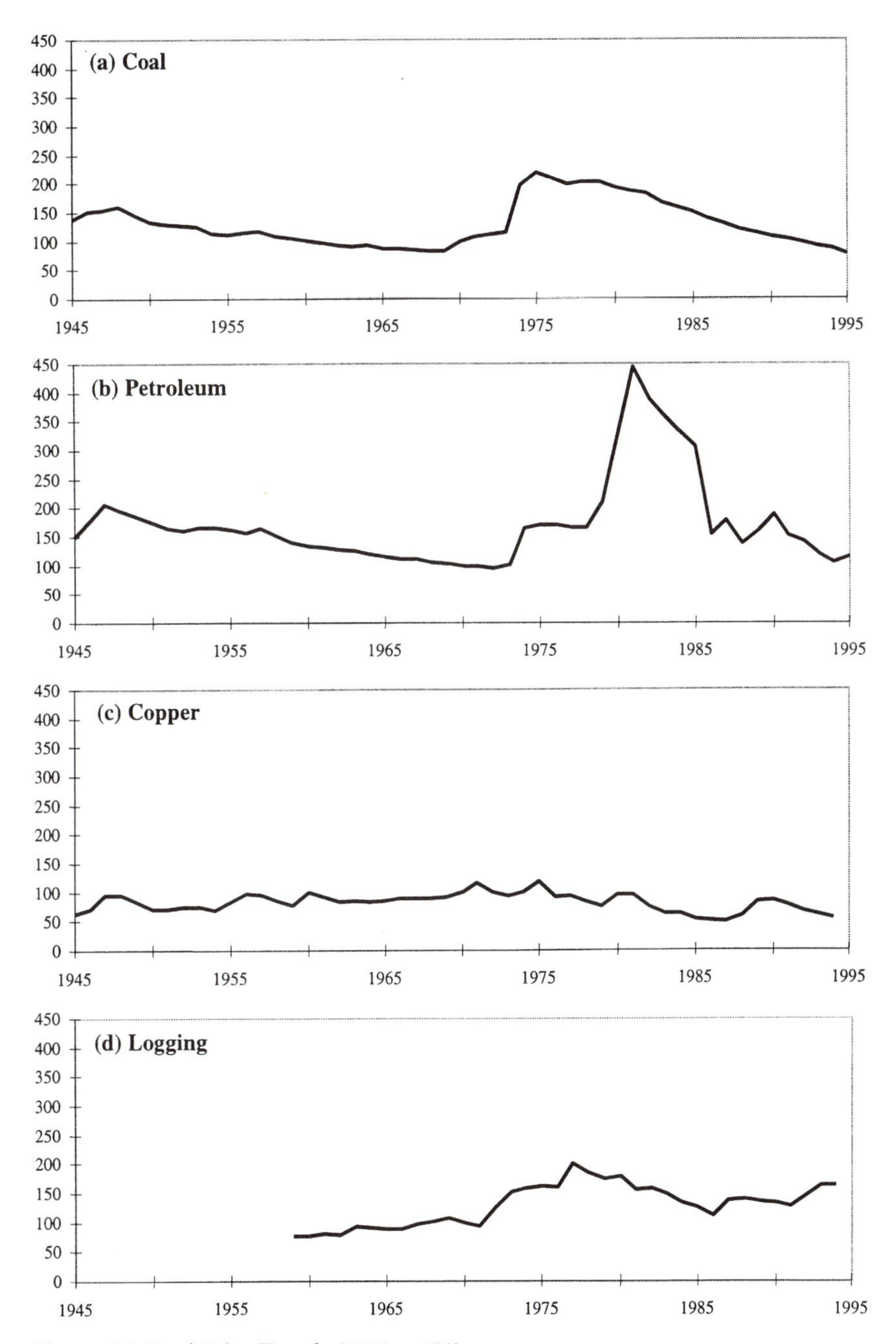

Figure 6-2. Real Price Trends (1970 = 100).

U.S. economy as a whole. During this period, output more than doubled in both industries. Coal output declined in the 1950s, in response to falling demand. In particular, diesel trains replaced steam trains, and electricity produced with oil and gas began to compete with coal-fired power plants. During the 1960s, however, the continuing expansion in demand for energy enabled coal output to recover. Logging output was only increasing slowly from the late 1950s to 1970, partly because of increasing competition from Canadian imports.

Coal and petroleum prices were declining steadily during 1945–1970. This suggests that any increase in demand in these industries was being more than offset by increases in supply. The main sources of supply increases were the opening up of new reserves and increases in extraction rates because of technological progress. Real copper and logging prices were on an upward trend during the 1950s and 1960s, although the increase in prices was very slight.[30]

Output trends in coal and copper mining have traditionally been somewhat erratic. One reason for this is that they have been prone to short-run disruptions in output from strike activity. Since the early 1980s however, these industries have become much more capital intensive, and the remaining labor is less unionized. As a consequence, there has been a dramatic fall in strike activity.

1970–1995. The experience in these industries during the twenty-five years after 1970 was very different from that in the twenty-five years before 1970. During the early 1970s output growth in all four industries stagnated but thereafter followed widely divergent paths. Coal and logging output expanded rapidly, ending the period 70% and 50% higher, respectively, than in 1970. There has been a slight contraction in logging since the mid 1980s, because new environmental regulations have limited access to forestlands. The copper industry declined after 1970, and by 1982 output had fallen by more than 30%. However, the industry then made a strong recovery, and output almost doubled over the next thirteen years. Unlike the other industries, petroleum output has fallen monotonically over the last twenty-five years and is now around 20% below peak output in 1970.

Prices increased in all the industries during the 1970s but then fell back in the 1980s. However, the extent of these price movements differed substantially among the industries. The most dramatic was petroleum, where the relative price increased by 350% from 1970 to 1981, then collapsed, and by 1994 was back down to 1970 levels. Coal prices also rose initially—by more than 100%—but then decreased steadily from the mid-1970s onward and were around 20% below 1970 levels by 1994. Copper prices followed a number of cycles, though they were on a downward trend and ended the period 45% lower. Logging prices doubled between 1970 and 1977, then fell back sharply, only to increase again after 1986. By the end of the period, logging prices were around 60% higher than 1970 levels.

As is well-known, the U.S. petroleum price trends during this period are mainly explained by developments in the world oil market. World oil prices rose dramatically in 1974 and again in 1979, following reductions in supply from members of the Organization of Petroleum Exporting Countries. Prices then fell back as these countries began to exceed their agreed quotas, and oil importers substituted for other fuels and developed more energy-efficient technologies. What is surprising is that these changes had very little impact on U.S. petroleum production.[31] As discussed in Chapter 3 of this book, there was a marked increase in exploration activity up to the early 1980s, but this was offset because the average rate of discovery for new oil wells fell significantly. In addition, previously known but unprofitable wells were developed, but this was offset by declining production as some existing wells were exhausted. After the early 1980s, exploration activity decreased and some higher-cost wells were shut down. However, the reduction in production was partially offset by increases in efficiency due to technological developments such as horizontal drilling and deepwater production platforms (see Chapter 3).

The increase in oil price led to an increase in demand for other fuels, especially coal, in the 1970s. Hence coal production, and particularly prices, rose rapidly. The continued expansion in coal output over the last fifteen years is surprising at first glance. This is because the demand curve for coal was "shifting in" as the price of oil fell, and environmental regulations were imposed on (downstream) coal-fired power plants. However, this effect was more than offset by a downward shift in the supply curve, caused by rapid productivity growth (see below).

Logging prices followed the same pattern of rising and falling over the period. The same is true for most natural resource prices, including land. A possible explanation for this is that the general climate of high inflation in the 1970s led to a shift out of financial assets and into real assets such as land, which bid up the prices of natural resources on the land. The reverse happened in the 1980s as inflation was brought under control.[32] The slowdown in output growth and increase in logging prices since the mid-1980s seem to be caused by new environmental regulations restricting logging, particularly in areas of natural habitat on public lands (see Chapter 5).

Copper prices followed the same downward trend in the 1980s, but unlike other natural resources, copper prices did not rise substantially during the 1970s. Prices in the 1970s may have been influenced by unique developments in the world copper market. In particular, the Chilean copper industry became more competitive in the 1970s following privatization and overtook the United States as the world's largest exporter. The fall in copper prices induced a substantial restructuring of the U.S. industry away from small-scale mines toward larger-scale, more efficient mines. Productivity growth in this much leaner industry accounts for the expanding output since the early 1980s (see Chapter 4).

Shares in Value of Manufacturing Output

To put some perspective on these trends, Table 6-1 illustrates how the shares of these industries in the total value of manufacturing output have changed over time. The petroleum share fell sharply from 16.9% to 7.9% between 1955 and 1965, mainly because of the fall in price of petroleum relative to other manufacturing goods. Despite rebounding during the high prices of the 1980s, the petroleum share has now fallen to 3.4%, reflecting the absolute decline in output below 1970 levels. During the period of high coal prices, the share of coal output in the value of manufacturing output rose to 4.1% in 1975. This has since halved—in spite of much higher output—because of the decline in relative coal prices. In contrast, the share of logging and copper output in the value of manufacturing output have been increasing over the last decade. However, in the case of copper, the share is still below its level prior to the restructuring of the industry in the early 1980s.

Labor Productivity

In the Four Industries. We begin our examination of productivity trends by looking at how labor productivity, Q_t/L_t, has changed over time in the four industries. This typically overstates multifactor productivity growth, because capital and intermediate goods tend to grow faster than labor.[33] However, since single- and multifactor measures of productivity generally move in the same direction, it gives an indication of qualitative trends in productivity. Moreover, it makes for consistent comparisons with simple labor productivity measures for the rest of the economy and for earlier periods within the four industries.[34]

Table 6-1. Share of Industry Output in Value of Manufacturing.

Industry	*1955*	*1965*	*1975*	*1985*	*1994*
Coal	3.1	1.2	4.1	3.3	2.0
Petroleum	16.9	7.9	7.9	11.6	3.4
Copper	0.6	0.5	0.5	0.2	0.4
Logging	1.1	0.7	1.2	1.3	1.7

Figure 6-3 shows labor productivity in all four industries for the period 1945–1995. Beginning with the latter twenty-five years, labor productivity is generally stagnant or declining until the mid-1970s and thereafter follows markedly different patterns. In the coal industry, labor productivity more than doubled between 1970 and 1994 and grew at an annual average rate of 2.4%. This is even more remarkable, given that labor productivity fell significantly during the 1970s and was still 25% lower in 1978 than in 1970. Indeed, over the last fifteen years, output increased by 24% while labor input fell by 52%! Labor productivity growth was even stronger in the copper industry, finishing the period 170% higher than in 1970. The average annual growth in labor productivity was 3.7%. In the petroleum industry, labor productivity fell substantially, to only 40% of its 1970 level by 1982. Since then, it has recovered somewhat but was still 35% below the 1970 level in 1994. In logging, labor productivity recovered strongly in the 1980s and was 60% above the 1970 level by 1986. However, it has since fallen back slightly and leveled off. The average annual growth in labor productivity over the whole period was approximately –2.4% in petroleum and 1.4% in forestry.

In the twenty-five years prior to 1970, labor productivity increased steadily in all four industries, with comparatively minor fluctuations about this trend (the exception to this is logging, where labor productivity growth was more sluggish). Thus, the productivity slowdown in the 1970s and the widely differing trends thereafter are in marked contrast to the experience in the first half of the post-1945 period.

In Broader Sectors of the Economy. Figure 6-4 indicates labor productivity trends in broader sectors of the economy—gross domestic product and manufacturing—over the same period. Again, in both these sectors, labor productivity growth increased steadily for the first twenty-five years, stopped in the 1970s, but then recovered to end up roughly 20% higher than 1970 levels by 1994. However, the productivity growth rate achieved in the last fifteen years (determined by the slope of the curves in Figure 6-4) has not been as high as that from 1945 to 1970. These trends are qualitatively very similar to those in the industries studied here, although the recovery in coal, copper, and logging was relatively stronger and in petroleum was relatively weaker.

Multifactor Productivity

Figure 6-5 shows multifactor productivity trends in the four industries, that is output divided by total factor input at time t, or $Q_t / (\pi_t^L L_t + \pi_t^K K_t + \pi_t^M M_t)$, where the 1970 value is normalized to 100. For petroleum, the multifactor productivity trend is very similar to the labor productivity

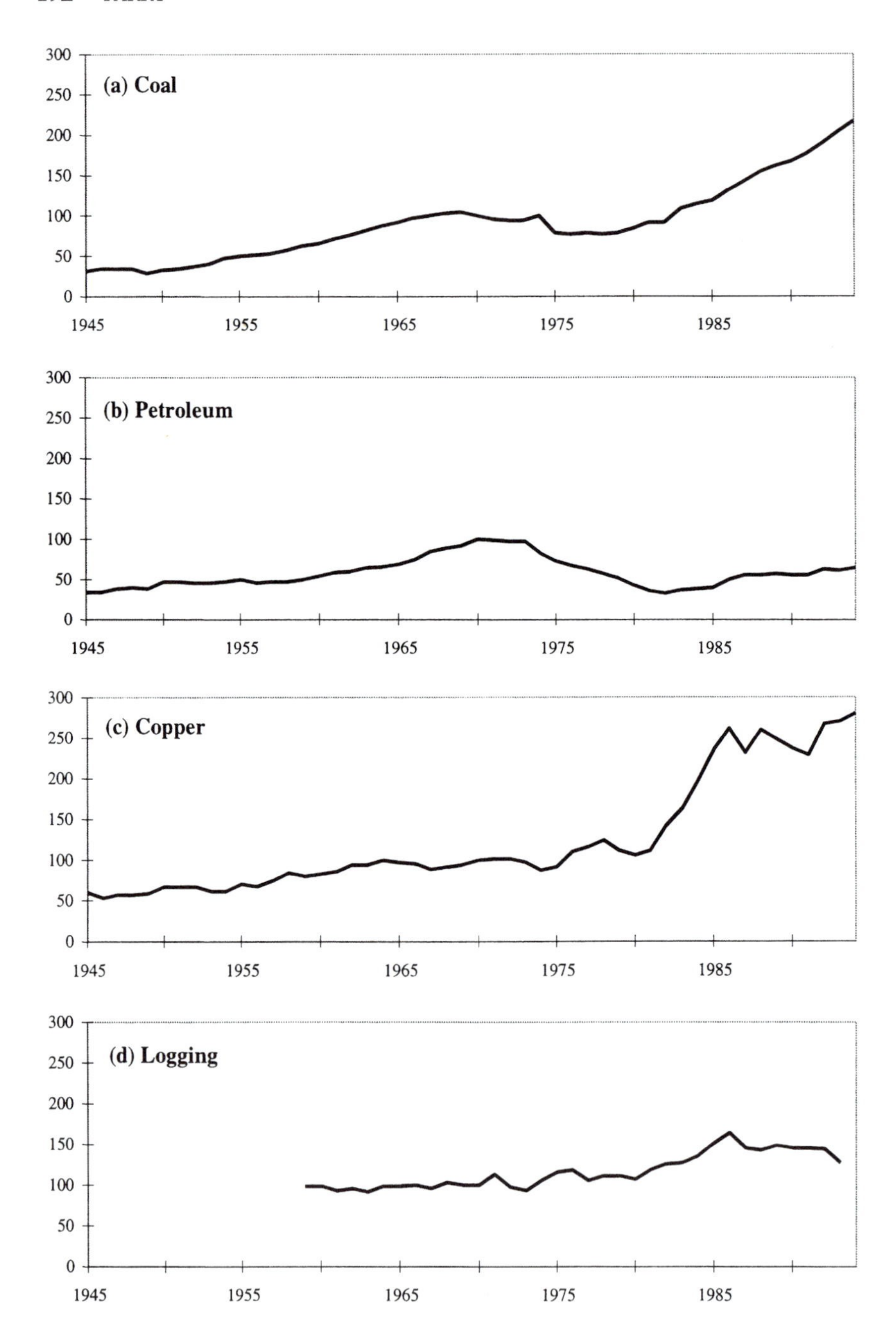

Figure 6-3. Labor Productivity Trends (1970 = 100).

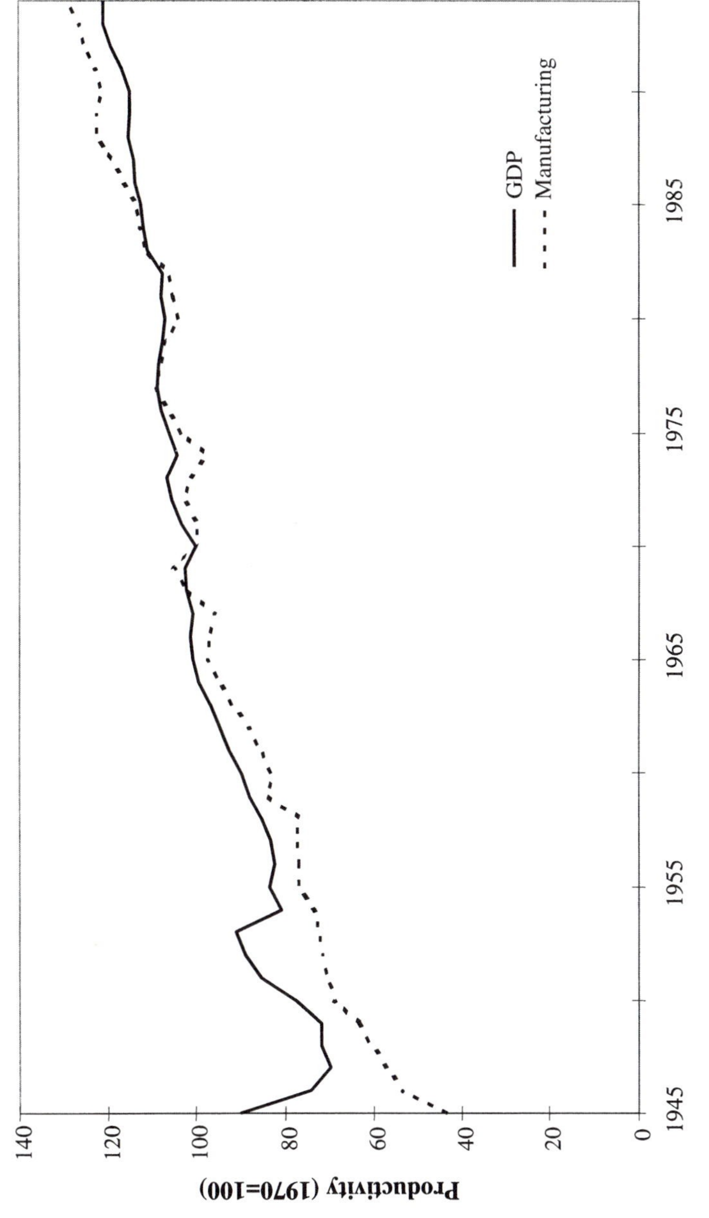

Figure 6-4. Economywide Trends in Labor Productivity (1970 = 100).

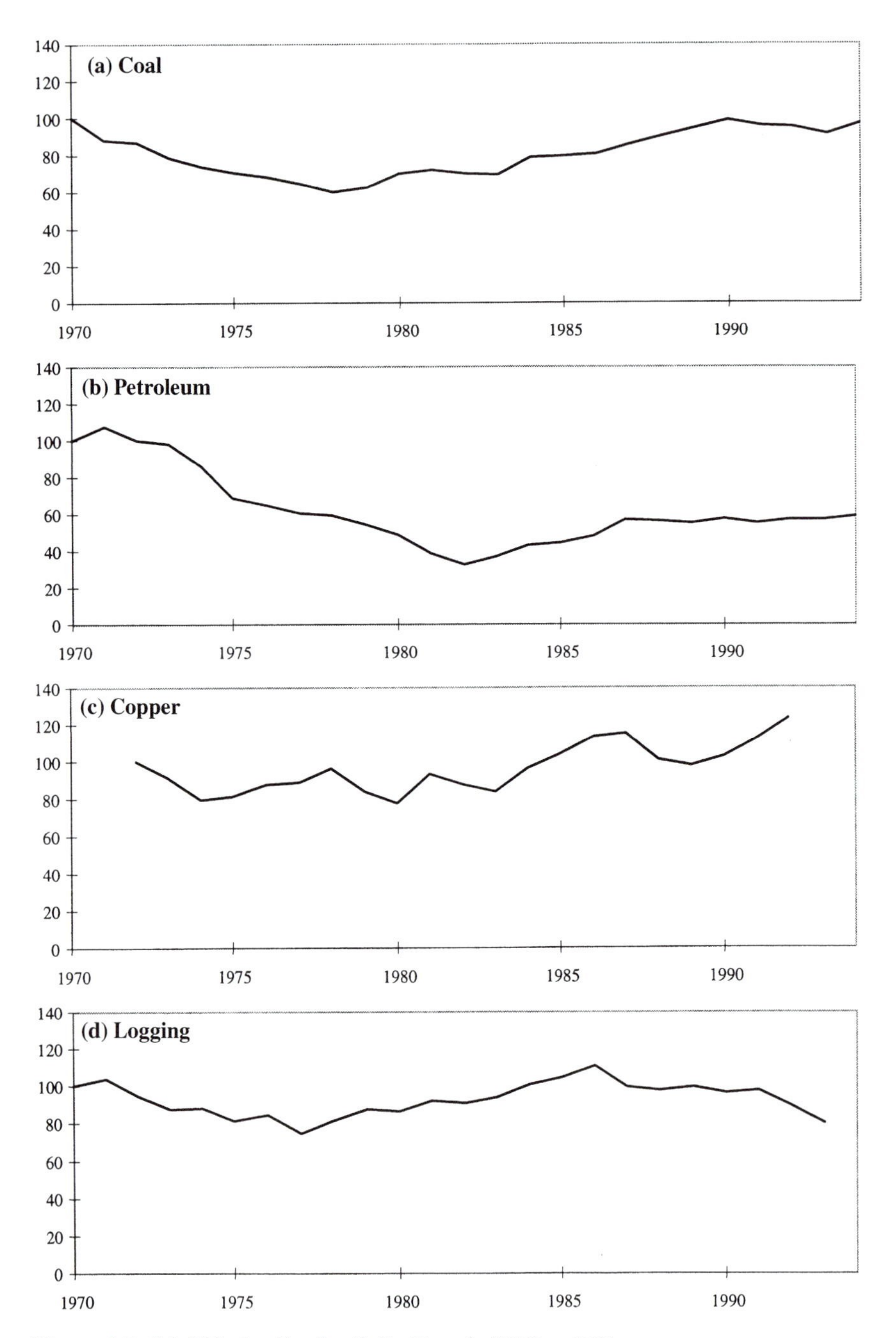

Figure 6-5. Multifactor Productivity Trends (1970 = 100).

trend. That is, productivity declines steadily to only 33% of the 1970 level by the early 1980s! However, productivity is now 75% higher than in the trough of the productivity fall. The productivity recovery in the coal industry during the 1980s is drastically lower than implied by the labor productivity measure. By the end of the period, multifactor productivity is slightly below the 1970 level, compared with the 120% increase in labor productivity. However, compared with the late 1970s, productivity has increased by around 64%.[35] For copper, the multifactor measure still indicates a significant rise in productivity over the last fifteen years, although not as dramatically as implied by the single-factor measure. By 1992, productivity was 23% higher than at the start of the period and 60% higher than at the trough of the productivity slowdown in 1979. The productivity performance in logging is worse when the multifactor measure is used. There was a noticeable recovery in productivity during the 1980s, although this has been reversed somewhat in recent years. Productivity is now 20% higher than in the trough year, 1977.

Multifactor productivity in manufacturing as a whole followed a similar qualitative pattern during the period. However, the decline during the 1970s was much more severe in the natural resource industries, suggesting that there were special factors at work (see below). Since the early 1980s multifactor productivity in manufacturing has increased by around 20% and is around 28% higher than in 1970.[36]

Contribution of Inputs to Output Growth

Figure 6-6 shows the contribution of input quantities over time in "explaining" output trends. The dashed curves are, from bottom to top, the weighted capital input, weighted capital plus labor input, and weighted capital plus labor plus intermediate inputs, where the weights are the shares in value product, or π_t parameters. Thus the difference between these dashed curves indicates approximately (since the relative weights vary somewhat over time) how input quantities change. The gap between output (the solid curve) and the top dashed curve is an absolute measure of productivity change between a particular year and 1970. When this dashed curve lies below (above) the solid curve, productivity is greater (less) than in 1970.[37]

In the coal industry, inputs were increasing faster than output in the 1970s, hence productivity was declining. As discussed in Chapter 2, these trends can be broadly attributed to three factors. First was the entry into the industry of small, relatively inefficient mines, in response to the jump in coal prices. This increased output and reduced the average level of productivity. Second, a variety of health and safety standards and environmental regulations were introduced during this period. For example, in

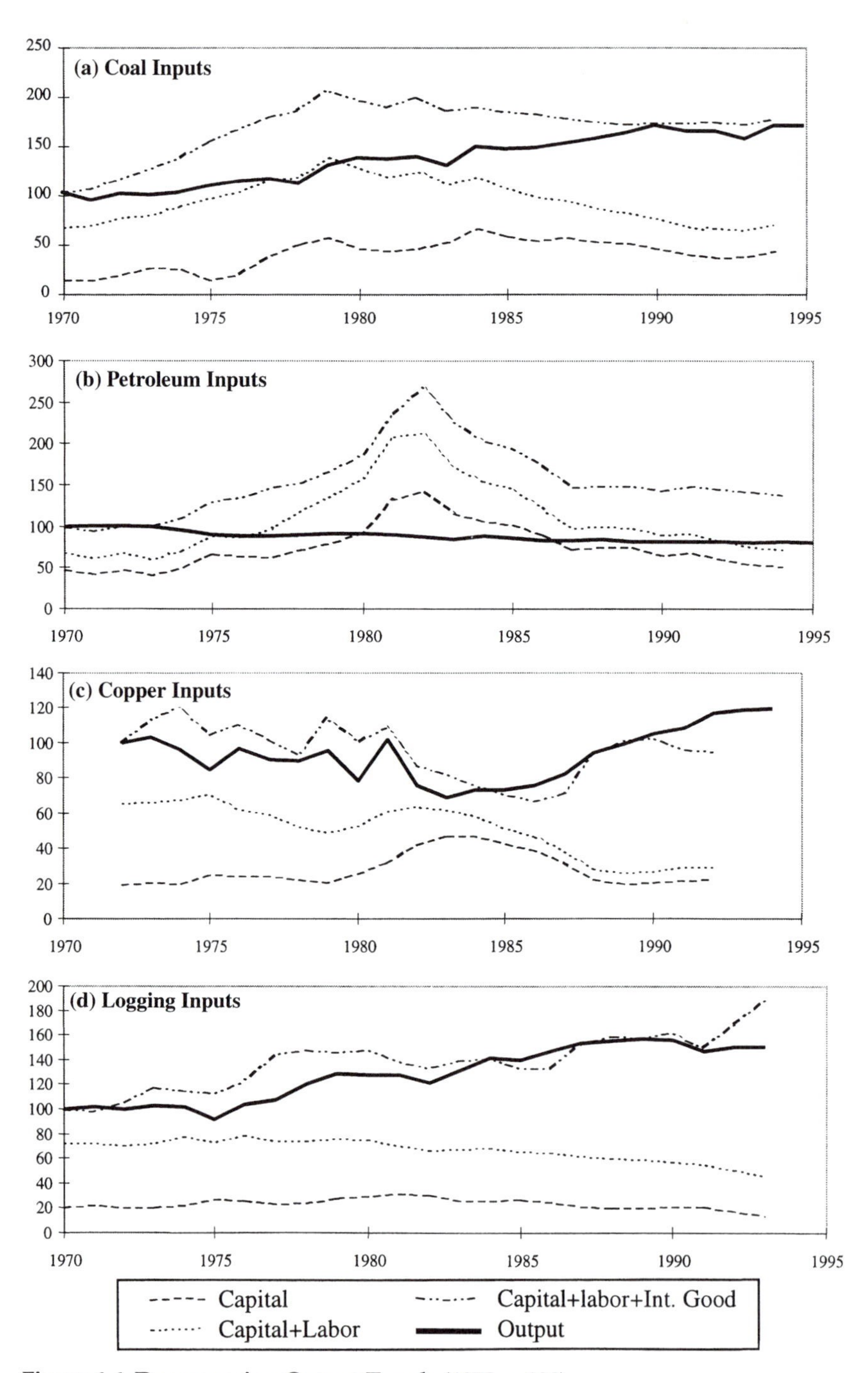

Figure 6-6. Decomposing Output Trends (1970 = 100).

Note: Input quantities are weighted by the share in value product.

underground mines, standards were established for roof building, ventilation, and allowable concentrations of coal dust and methane. Other regulations restricted mining in areas in close proximity to national parks and required that exhausted surface mines (on federal lands) be transformed back into the pre-existing vegetative state. Loosely speaking, such regulations have a once-and-for-all effect on increasing input requirements per unit of coal output and therefore have a permanent effect on the *level* of productivity but only a transitory effect on the rate of productivity *growth*. Third, this was a period of industrial confrontation, which culminated in the highly disruptive strikes of 1971 and 1977. Strikes have a once-and-for-all effect on reducing output and productivity below their potential. This climate of high prices and industrial strife during the 1970s most likely slowed down the adoption of new technologies.

In the 1980s, these factors were largely reversed. Falling coal prices drove many of the small-scale, less-efficient mines out of the industry, thereby boosting the average level of productivity. In addition, falling prices forced the industry to become more competitive. Productivity was increased by the adoption of labor-saving technologies, such as longwall mining in underground mines and cutting by draglines in surface mines (see Chapter 2).

Similarly, in the petroleum industry, inputs increased sharply in the late 1970s and early 1980s. This had no noticeable effect on output, hence productivity fell markedly. As mentioned above, the increase in petroleum price led to a big increase in exploration activity (the "extensive margin") and the development of known wells that were previously too costly to be profitable.[38] The fall in average productivity was compounded by the depletion of the "intensive margin," that is, the exhaustion of reserves in lower-cost wells.[39] In the 1980s, the fall in petroleum price led to the closure of many high-cost wells (that is, depletion of the extensive margin); thus the quantity of capital was reduced and productivity rose. Technological breakthroughs further increased productivity and mitigated the fall in output (see Chapter 3). These included the development of floating platforms for petroleum extraction in deepwater areas and techniques to drill horizontally into petroleum deposits. However, in the 1980s, there was also an important trend toward contracting out activities, such as site-surveying and well-drilling, rather than providing them from within the firm. These activities have become much more high-tech, but this is not reflected in a productivity improvement in Figure 6-6, because they are now intermediate inputs in production.

In the copper industry, the quantity of inputs and output produced were (roughly) on a constant trend during the 1970s. Then, in the early 1980s, the quantity of labor and capital were reduced sharply, and thereafter productivity rose steadily. As discussed in Chapter 4, the story of the copper industry over the last twenty-five years is one of remarkable recovery from being on the brink of collapse. By the end of the 1970s, the industry had become highly uncompetitive. For whatever reasons—opposition from unions or timid management—falling product prices had not led to the closure of loss-making mines or reductions in wages.[40] Then, in the early 1980s, everything changed and the whole industry was restructured. Many jobs were lost (for example, hours worked in 1986 were only one-third of those in 1981), and average real wages fell by 25% over these five years. This led to a dramatic shake-out of small-scale, less-efficient mines and consolidation into a smaller number of larger, more efficient mines. In addition, there has been significant technological innovation over the last fifteen years, for example, the development of new chemical processes for extracting copper from copper ore (see Chapter 4).

The most noticeable feature of the panel for logging in Figure 6-6 is the increasing importance of intermediate goods over the last twenty-five years. This reflects an increase in the costs of road-building services as logging has shifted toward less-accessible areas. In part, this is because of a depletion effect as first-growth and second-growth forests are logged. But environmental regulations restricting areas that can be logged have also played a role, particularly since the mid-1980s. This depletion effect has limited overall productivity growth, in spite of some technological improvements in logging, such as the replacement of the chainsaw by mechanical tree fellers in certain regions.

In fact, the forestry industry more broadly is in the middle of a transition away from logging in virgin forests toward a sustainable "agricultural" industry based on tree planting, growing, and harvesting. As discussed in Chapter 5, this trend is partly a response to increasing environmental regulations limiting access to traditional forest areas. In addition, the attractiveness of tree farming has been increasing because of improvements in productivity. These have included the selection of superior species for planting and, more recently, the development of genetically engineered trees. However, these important improvements in the forestry sector more broadly do not show up in Figure 6-6, because this focuses purely on harvesting and not tree growing.

Summary Statistics

Table 6-2 summarizes the average annual rate of growth in multifactor productivity in the industries for 1970–1980 and for 1980–1992. During the 1970s, this ranged from –1.5% in logging to –7.0% in petroleum. For manufacturing as a whole, multifactor productivity growth averaged 0.8% per year. This was well below historical levels, but significantly better than the performance in the natural resource industries. During 1980–1992, productivity growth in the natural resource industries ranged

Table 6-2. Summary Statistics on Average Annual Growth in Multifactor Productivity.

	1970–1980	*1980–1992*
Coal	–3.5	2.6
Petroleum	–7.0	1.3
Copper	–2.5	3.9
Logging	–1.5	0.3
Manufacturing	0.8	1.4

from 0.3% in logging to 3.9% in copper. For manufacturing, productivity growth was 1.4% per year over this period. Broadly speaking, the performance of the natural resource industries since 1980 has been more in tune with—if not better than—that for manufacturing as a whole.

Many explanations have been suggested for the decline in manufacturing productivity growth during the 1970s. These include: the increase in price of energy following the oil price shock; a reduction in the rate of savings and investment; a slowdown in the accumulation of knowledge from R&D activity; the effect of new regulations such as the Clean Air and Clean Water Acts; and the crowding out of private industry by the increasing share of public spending in gross domestic product. Since the productivity decline in the natural resource industries was initially much more severe, this suggests that additional factors were at work in the natural resource industries. One such factor was the cushioning of the natural resource industries in the 1970s due to the unusually high product prices. This allowed relatively inefficient producers to enter these industries. In addition, the lack of competitive pressure reduced the incentive for technological innovation and to avoid disruptive labor disputes.

SUMMARY

This chapter estimates multifactor productivity trends in four natural resource industries—petroleum, coal, copper, and logging—over the last twenty-five years. During the 1970s, productivity declined in all four industries. This contrasted sharply with the historical experience of gradually declining average production costs in natural resource industries identified by Barnett and Morse (1963). At the time, this led to concern that natural resource endowments were being exhausted and that this would limit future opportunities for economic growth.

In hindsight, the 1970s appear to be an exceptional period rather than marking a sea change in long-run productivity trends. During that decade, unusually high natural resource prices encouraged the entry of relatively inefficient producers, particularly in the coal and petroleum industries, as discussed in Chapters 2 and 3. In the coal and copper industries, high product prices also seem to have strengthened union opposition to new technologies and to the closure of less-efficient mines. In addition, the phasing-in of new environmental and health and safety regulations had a once-and-for-all impact on depressing productivity levels, notably in the coal industry (Chapter 2). Since the early 1980s, productivity has been increasing in all four industries. Part of this appears to be due to restructuring and downsizing in response to lower output prices. The most dramatic example of this was the shakeout in the copper industry (see Chapter 4). However, technological improvements, such as longwall coal mining and horizontal drilling in petroleum, have also played an important role. The productivity recovery is less pronounced in the logging industry, due to the exhaustion of first- and second-growth forests and increasing restrictions on logging for environmental reasons (see Chapter 5).

ACKNOWLEDGMENTS

I am very grateful to Charles Hulten and Larry Rosenblum for extremely valuable comments on an earlier draft of this chapter. John Anderson, Douglas Bohi, Joel Darmstadter, Hans Landsberg, Paul Portney, Roger Sedjo, Dave Simpson, John Tilton, Mike Toman, and two referees also provided helpful suggestions. I thank Doug Harris and Brian Kropp for outstanding research assistance and the Sloan Foundation (grant number 96-3-2) for financial support.

ENDNOTES

1. For a discussion of possible reasons for the productivity slowdown see the symposium in the Fall 1988 edition of the *Journal of Economic Perspectives*.

2. The most well-known study was by Meadows and others (1972), which was prepared for the Club of Rome.

3. For lucid critiques of these studies, see Nordhaus 1992 and Simon 1996.

4. In this series of studies, we will not be concerned with economic efficiency per se. Thus we do not attempt to quantify possible market imperfections caused by environmental externalities, imperfect property rights, government subsidies, and so on. (For a technical discussion of these issues, see Stiglitz 1979.)

5. For example, practically all previous studies of productivity in forestry focus on processing rather than harvesting (see Stier and Bengston 1992). In addition, our purpose is to look for common patterns across industries, while most other studies have focused purely on one industry.

6. See, for example, Anderson 1977 for a detailed discussion.

7. See, for example, Denison 1962, 1967, 1974; Kendrick 1961, 1973; Jorgenson and Griliches 1967; Griliches 1960; Jorgenson and others 1987; and Kendrick and Vaccara 1980. This approach has been more common than econometric analyses in time series studies of productivity. The reason is that input and output quantities are closely correlated over time, making the individual regression coefficients difficult to decompose.

8. Purchased services are activities performed by outside contractors rather than within the firm. Examples may include exploration and drilling activities in petroleum and road building in logging.

9. Improvements in the quality of the labor force—such as increases in the average level of education, skill, and health of employees—affect the level of productivity. However, these appear to have been less important during the period than the other factors.

10. Competition can be consistent with decreasing returns to scale at the firm level. Rents accrue to the scarce input—the resource stock. Part of the rent goes to private owners of resources. The rest goes to the government when it sells exploration and development rights, or imposes royalty fees.

11. If we had not assumed competition and therefore been unable to use Eq. 6-2, the πs would be the elasticities of output with respect to each factor input. However, data on these elasticities are not available, whereas the factor shares can be measured.

12. We did calculate productivity growth for each industry using the real input price as a proxy for input quality. However, this adjustment had very little impact on reducing the productivity residuals. For more discussion about measuring changes in input quality, see, for example, Jorgenson and others 1987.

13. For more detailed discussions of the methodological issues, see, for example, Denison 1974; Jorgenson and Griliches 1967. All the raw data used are available from the author upon request.

14. This includes wood harvested for both lumber and pulp and paper production.

15. Indeed, measuring quality changes for the output of nonresource industries is problematic. Usually, quality is proxied by the real price of output. However, the price of output can also change due to other factors, such as shifts in demand.

16. It is not possible to derive productivity measures for oil and gas individually because the inputs into these industries cannot be separated out.

17. For a discussion of how to adjust conventional productivity measures to take into account environmental impacts, see Repetto and others 1996.

18. For nonproduction workers (accountants, payroll personnel, and so on), only the total number of employees is available. We assume that each of these employees works thirty-five hours per week for forty-nine weeks per year. The ratio of nonproduction workers to production workers is typically less than 0.1.

19. In particular, the current value of an asset may differ substantially from its original purchase price because of inflation (which was substantial during the 1970s). Furthermore, the formulas for depreciation used by accountants to calculate book value (such as straight-line depreciation) may bear little resemblance to true economic depreciation.

20. These price indexes (which are from the Bureau of Labor Statistics) are for industry definitions that are similar to those used above. For logging, we used producer price indexes for manufacturing as a whole.

21. These are estimated rates of depreciation averaged during 1970–1994. For coal, these are 0.17 (equipment) and 0.08 (structures); for petroleum, 0.15 and 0.115; and for copper, 0.15 and 0.08. Estimates of capital depreciation rates in logging are not available. We assumed values of 0.15 and 0.08. I thank Larry Rosenblum for supplying these data. The capital stock series are not particularly sensitive to alternative plausible values for depreciation rates.

22. Given our values for depreciation, this implies that only around 2% and 25% of the 1945 book value of equipment and structures, respectively, is left by 1970. For copper and logging, because of data limitations, we had to use a more recent value of K_{i0}.

23. We use the six-month Treasury bill rate (obtained from the *Economic Report of the President*). To convert the Treasury bill rate into a real interest rate, we subtract the rate of consumer price inflation. The capital stock series we obtain are not sensitive to using alternative interest rates, including those adjusted for taxes.

24. During the late 1970s and 1980s, there was a shift in investment away from copper and other metals towards gold. As a result, there may be some upward bias in our estimates of the growth of capital in the copper industry, and hence some underestimate of the growth in multifactor productivity.

25. Again the exception is logging, and for this industry we used price indexes for manufacturing as a whole.

26. However, the overall productivity trends for the industries turn out to be similar under this approach to those presented below.

27. Changes in output prices imply substantial changes in the sum of the input weights over time. This would result in our estimated productivity series being distorted by output price effects. To avoid this problem, we scale the input weights such that they sum to the same value in all periods. This value is the average sum of the weights over the twenty-five-year period. However the *relative* input weights are variable over time. The weight attached to labor is slightly biased downward because fringe benefits and social security taxes paid by employers are not included in labor costs. The capital weight is also slightly understated because estimated capital costs do not include corporate tax payments. These payments are not as large as they may appear because interest payments, depreciation, and (at times) investment expenditures are tax deductible.

28. Data for purchased services are only available for two of the twenty-five years. However, in these years, it is only around 5% of intermediate goods expenditures, and therefore we ignore purchased services for this industry.

29. Data for logging are only available from 1958. For earlier periods, only broader measures of logging output are available.

30. A price index for logging was obtained by dividing the value of output by the total volume of wood harvested.

31. However, to some extent these figures mask an increasing share of natural gas relative to oil in total petroleum production.

32. Consistent with this explanation, real stock prices fell in the 1970s and rose sharply in the 1980s.

33. That is, output divided by (weighted) total factor input grows more slowly than output divided by labor input.

34. Estimates of multifactor productivity for the four industries in earlier periods are more difficult to obtain because of data limitations.

35. These trends are consistent with a recent study of multifactor productivity in the coal industry by Ellerman and Berndt (1997).

36. These data are from the Bureau of Labor Statistics website, http://stats.bls.gov/datahome.htm (accessed January 1998).

37. The curves in Figure 6-5 are equivalent to the ratio of the solid curve in Figure 6-6, divided by the highest dashed curves.

38. This effect was stronger because petroleum price controls in the United States were lifted during the 1970s. These had artificially inflated the level of productivity by keeping marginal, relatively less-efficient producers out of the industry.

39. In contrast, the depletion effect has not been significant in coal because reserves are very plentiful (see Chapter 2).

40. In addition, the government turned down petitions in 1978 and 1984 for protecting the industry against foreign imports.

REFERENCES

Anderson, L.G. 1977. *The Economics of Fisheries Management*. Baltimore: Johns Hopkins University Press.

Barnett, H.J., and C. Morse. 1963. *Scarcity and Growth: The Economics of Natural Resource Availability*. Baltimore: Johns Hopkins University Press.

Denison, E.F. 1962. *The Sources of Economic Growth in the United States and the Alternatives Before Us*. New York: Committee for Economic Development.

———. 1967. *Why Western Growth Rates Differ: Postwar Experience in Nine Western Countries*. Washington, D.C.: Brookings Institution.

———. 1974. *Accounting for United States Economic Growth, 1929–1969*. Washington, D.C.: Brookings Institution.

Ellerman, D.A., and E.R. Berndt. 1997. An Initial Analysis of Productivity Trends in the American Coal Industry. Unpublished manuscript, Massachusetts Institute of Technology.

Griliches, Z. 1960. Measuring Inputs in Agriculture: a Critical Survey. *Journal of Farm Economics* 42: 1411–27.

Hulten, C.R. 1992. Accounting for the Wealth of Nations: The Net versus Gross Output Controversy and Its Ramifications. *Scandinavian Journal of Economics* 94(supplement): 9–24.

Jorgenson, D.W., and Z. Griliches. 1967. The Explanation of Productivity Growth. *Review of Economic Studies* 34: 249–83.

Jorgenson, D.W., F.M. Gollop, and B.M Fraumeni. 1987. *Productivity and U.S. Economic Growth.* Cambridge, Massachusetts: Harvard University Press.

Kendrick, J.W. 1961. *Productivity Trends in the United States.* Princeton: Princeton University Press.

———. 1973. *Postwar Productivity Trends in the United States, 1948–1969.* New York: National Bureau of Economic Research.

Kendrick, J.W., and B. Vaccara, editors. 1980. *New Developments in Productivity Measurement and Analysis.* Studies in Income and Wealth, vol. 44. Chicago: University of Chicago Press for the National Bureau of Economic Research.

Meadows, D.H., D.L. Meadows, J. Randers, and W.W. Behrens. 1972. *The Limits to Growth.* New York: Universe Books.

Nordhaus, W.D. 1992. Lethal Model 2: The Limits to Growth Revisited. *Brookings Papers on Economic Activity* 2: 1–59.

Repetto, R., D. Rothman, P. Faeth, and D. Austin. 1996. *Has Environmental Protection Really Reduced Productivity Growth?* Washington, D.C.: World Resources Institute.

Simon, J.L. 1996. *The Ultimate Resource* 2. Princeton: Princeton University Press.

Stier, J.C., and D.N. Bengston. 1992. Technical Change in the North American Forestry Sector: A Review. *Forest Science* 38: 134–59.

Stiglitz, J.E. 1979. A Neoclassical Analysis of the Economics of Natural Resources. In *Scarcity and Growth Reconsidered,* edited by V.K. Smith. Washington, D.C.: Resources for the Future.

Index